# ELECTROLYTES
## (Trace Minerals)
# *THE SPARK OF LIFE*
### The Key to Longevity
### & Quality of Life

## Gillian V. Martlew, N.D.

Nature's Publishing, Ltd.
P.O. Box 361
Murdock, Florida 33938-0361

First Edition 1994
Printed in the United States of America

ISBN 0-9640539-0-X

Library of Congress Catalog Card Number
94-065134 Electrolytes, The Spark of Life

Other books by Gillian Martlew, N.D.
and Shelley Silver

The Medicine Chest
The Pill Protection Plan
Stay Well in Winter
The Green Home Handbook

It is the premise of this book that trace minerals are an
important key in the prevention of disease. And I stand by that,
whether or not the prevailing medical establishment agrees. I
also want to make the point that the material contained in this
book is meant for educational purposes only and is in no way
intended as a substitute for consultation with your physician or
health care practitioner. Neither the author nor the publishers
accept any liability, however arising, from the use of any of the
information contained herein, by any person whatsoever.

*"Unless we put medical freedom into the Constitution, the time will
come when medicine will organize itself into an undercover dictatorship.
To restrict the art of healing to one class of men and deny equal privileges
to others will constitute the Bastille of medical science. All such laws are
un-American and despotic."*
⚜
Dr. Benjamin Rush
Physician and signer of the
Declaration of Independence, 1776.

Published by Nature's Publishing Ltd.
P.O. Box 361. Murdock, FL 33938-0361

Printed by Rose Printing

i

This book is dedicated to Karin

And to my friends
Catherine and David
Pat and Kim
Jane
Mark and Alisa

And to the memory of
Cindy Buso

"All day Monday and until Tuesday noon,
creation was busy getting the earth going.
Life began Tuesday noon and the beautiful, whole, organic nature
of it developed over the next four days. At 4.00 P.M. Saturday the
big reptiles came on. Five hours later when the redwoods appeared
there were no more big reptiles. At three minutes before midnight,
man appeared. At one-fortieth of a second before midnight the
industrial revolution began.

We are surrounded with people who think that what we have been
doing for that one fortieth of a second can go on indefinitely.
They are considered normal, but they are stark raving mad."

David Brower

# CONTENTS

# ACKNOWLEDGMENTS

This book owes its existence to Dr. Gerald Olarsch N.D. It was his concept, and many of the new ideas I have explored are also his. Jerry is a retired Naturopath and devotes his time to the study of natural healing and the trace minerals. He has an extensive knowledge of natural therapies and a comprehensive health library of old and new books and research papers. It was through him sharing this treasure trove and his knowledge with me, and my work with patients over the years, that I was able to write this book. And for all this I am extremely grateful. I am lucky to count him as a friend and my Naturopathic mentor. Thanks Jerry.

Joan Olarsch is an incredible lady with a talent for doing five things at once and keeping her hair on. She runs a home, a business, and coordinated, faxed and mailed research material to me for more than a year. There is always the danger that the people who work away in the background get ignored. I noticed, and I appreciate all your effort on behalf of this book. Thanks Joan.

I want to thank Shelley Silver for all her hard work which helped to made the writing of this book possible. In addition to her usual invaluable help, she also read the book over and over to me so that while it was in the process of evolving I could hear what I was writing—and scribble things in the margins. What a ghastly job. Special thanks as always Shelley.

My mother always inherits the task of editing my books and restraining herself from putting thick red lines through all the split-infinitives. Thank you for wading through piles of manuscript for the fifth time. I also want to thank my parents for their unfailing enthusiasm for my projects, and their love. Thanks to two exceptional people.

David spent hours working with me on his Apple Macintosh computer, provided priceless technical information and shared his years of computer experience in order to help create the book's visual character. Glitches notwithstanding it was fun. Thanks David and 'Mac.'

Thanks to Karin for the cover idea, and Jim Davis for the cover design.

I would also like to acknowledge the work of Dr. Bernard Jensen and some of the great trailblazers in trace-mineral research; Dr. G.H. Earp-Thomas, William A. Albrecht, Dr. Henry Schroeder, Drs. Eric J. Underwood and Walter J. Mertz, John Hamaker, Donald Weaver, J.I. Rodale and Dr. Carl Pfeiffer.

# INTRODUCTION

## *Bernard Jensen, D.C., N.D., Ph.D.*

Electrolytes are vital to health, they are the building blocks of life. An electrolyte is a substance whose molecules split into electrically charged particles or ions when melted or dissolved. In this form, the ions then become capable of conducting electricity—an ionic solute. There are many ions that play an important role in regulating body processes. Some of these are potassium, sodium, hydrogen, calcium, magnesium, phosphate and chloride.

Most people do not realize the importance of electrolytes in their bodies. Electrolytes are the beginning of minerals in action. The point when minerals become alive is when electrolytes split. A transformation begins when minerals can become active and usable in human tissue. All cellular structures become alive through electrolytic activity. Life begins with electrolytes. Trace minerals carry the life force in our bodies more than any other substance.

The body is an electromagnetic organism. There is a vibrational emanation from every organ. These emanations have been tested with radionic instruments. We find that every organ has its own aura. Kirlian photography has been used to investigate the electrical vibrations and auras of the human organism. I feel that there has not been enough investigation into the finer forces of life in the research work of today. It is the electrical current that is given to us through the atmosphere that gives life to every plant, animal, and human on the face of the earth. We can notice this electrical current in the lightning, in thunder and in earthquakes.

It is time for people to realize and to understand that we are magnetic beings. We cannot live without the finer forces of life that come to us through electrolytes, through trace minerals, and through the vibrations of the atmosphere. We must get good clean water, fresh air, sunshine and nutrients from living plant life in order to bring the finer forces into our bodies. We must understand the importance of the soil. It must be rich in the elements that are necessary for the life and health of man so that the plants can absorb these elements and we can then nourish the cells of our bodies with them. We cannot rejuvenate our bodies properly without these finer elements that are so minute that they are even difficult to see under a high powered microscope.

In order to have healthy people on our planet, we must look to the deeper issues and the finer forces for answers. Besides good soil, we must have clean air so that the elements necessary for life can reach us without obstacle. With every breath we take in nourishment. In cities that are heavy with smog and pollution, people are breathing toxic wastes into their lungs. There is little space left in our polluted air for nourishing elements.

We are polluting our waters. Water is one of the finest mediums to bring us proper nutrients. We drink water directly as well as take it into our bodies through the juices of plants. Our plants need to be nurtured with good water in order to be healthy. Our plants need an adequate amount of fresh air and sunshine in the same way that we do. Vegetables and fruits that grow in polluted areas cannot receive the fresh air and sunshine necessary to provide us with the nourishment that we need.

When people get well through good water, nourishing foods, fresh air and sunshine, they may not realize the important role that the trace minerals and electrolytes played in their healing. It is in the electrolyte that the cosmic ray is formed. It is in electrolytes that molecular movement begins. It is through the work of electrolytes that we can see what nature really has in store for us. The green material and copper that is in chlorophyll are elements that keep us from becoming anemic. Electrolytes must be in balance within the body for certain good bacteria to exist and carry on their work of fighting harmful bacteria. It is because of electrolytes that our immune systems can function in good order.

Science has not begun to know everything about the work of electrolytes within human bodies or the bodies of plants and animals. Science often skirts around the finer forces of life. It was Milliken who said, "Where man ends, God begins." Electrolytes are responsible for transferring energy and for the regeneration and rejuvenation of every cell. A new healing art will soon be born that will recognize and utilize the power of these tiny powerhouses called electrolytes. From the moment the sperm is attracted to the egg in the body of the mother, the electromagnetic forces of electrolytes are at work. Bodies cannot be built without electrolytes.

Scientists are going to do more research in the area of

electrolytic activity. They will develop an instrument that will detect the magnetic effect of electrolytes in the body. It is the live body, life itself, that carries electrolytic activity. It is time for scientists to realize that there is something deeper in man that has to be cared for. We cannot heal a person through symptomatic relief. We must go to the root of the problem, to the cell itself—where electrolytes are at work giving life. It is here that real healing can begin.

*Dr. Bernard Jensen*

# PREFACE

*"The real voyage of discovery consists not in
seeking new landscapes, but in having new eyes."*

Marcel Proust

What you hold in your hands is the third rewrite of a book I thought would take a couple of months to complete. It didn't.

The first draft read like a fusty old school book, so I filed it in the garbage. And the second one took months to metamorphose into the third. Why am I telling you all this? Because I am excited by the way this book has evolved, and especially with what I have to share with you.

In the two decades that I have dedicated to natural healing there have been many moments of enlightenment, one of which was coming to understand the importance of electrolytes, or trace minerals, and their relation to health. I may have missed their significance had they not been brought to my attention by my friend and mentor Dr. Jerry Olarsch.

The information in this book has been under our noses for decades, but buried in research papers and forgotten. Pioneers like Dr. Earp-Thomas and William Albrecht were researching the soil, trace minerals and electrolytes, and their effect on health, way back in the first half of this century. But their work is only known to a handful of people.

Science and medicine have been too busy searching for patentable pharmaceuticals to have paid attention to research that won't make serious money. And so lifetimes of work on trace minerals and electrolytes have lain dormant for decades, waiting to be rediscovered.

At the risk of sounding pompous, I'm going to suggest that the information in this book breaks the 'rule' that everyone is different, and therefore no single piece of information applies to every person. I believe this absolutely in the context of health, orthodox and natural remedies—and life in general. But I equally believe that the significance of the material you are about to read affects every human being, every animal and every plant on this planet—even this planet. By the time you have finished reading this book you will have one more piece to fit into your own personal health puzzle. The *key* piece.

If everything ailing us was reduced to its lowest common denominator, aside from the individual considerations

*Lack of Electrolytes or Trace minerals Elements in every affliction*

and psychological aspects of each, every affliction would have, somewhere in its etiology, a lack of electrolytes or trace elements. These minute nutrients are involved at the fundamental level of every one of the body's functions.

Having said that, trace minerals should be part of a holistic approach to life and health. If we think that good nutrition alone will make us completely healthy we are in for a rude awakening. Likewise, it doesn't matter how spiritual we are, how much yoga, meditation or exercise we do, or how much we have learned to use the power of our minds. The body still needs the basic materials or it will never be totally healthy.

Electrolytes are the spark; the catalyst that makes everything in the body function. If they are in short supply and remain that way, the body begins to degenerate. Irrespective of whatever problem we are talking about; lack of energy, child hyperactivity, high cholesterol, heart disease, arthritis, depression, constipation, or even Alzheimer's disease—we always need to go back to *basics* to look for answers, and to begin the healing process. A Health program which includes a rebalancing of electrolytes creates the physical environment where healing and rejuvenation can happen, and the body begins to work for itself.

In this crazy world we've created where the body, mind and spirit are constantly subjected to things our ancestors couldn't even dream about; basic good nutrition has never been so important. We have gone against nature, depleted and poisoned the air, water, soil and food, and are reaping the negative pay-back as cancer, heart disease, AIDS and other degenerative and immune disorders.

I believe that trace minerals, especially in electrolyte form, are one of the most important and overlooked aspects of preventative and restorative health care of the twentieth century.

*Rebalance electrolytes creates physical environment where healing and rejuvination can happen, — and body begins to work for itself.!*

# ELECTROLYTES & MINERALS
# AN OVERVIEW

*"...minerals are now gaining some dignity
after a long and ignoble status as ash."*

Dr. Edward Howell

## THE SPARK OF LIFE

For thousands of years people have been aware of the potential of minerals, seeking out mineral waters and springs and believing in their healing and rejuvenating potential.

Trace minerals and electrolytes are the spark of life, and without them we simply wouldn't function. Minerals are the basic components of all matter. They are built into key enzymes and hormones, and are part of cells, tissue, bone, blood and body fluids. They also assist in every aspect of life from the production of hormones and energy, digestion, nerve transmission and muscle contraction, to the regulation of pH, metabolism, cholesterol, and blood sugar. Without minerals, protein (the body's building material), and vitamins cannot be utilized and have little or no function in the body.

Minerals and trace minerals supply neither energy nor fuel to the body but are instrumental in their production and use. Most of the chemical reactions in the body are governed and catalyzed by many thousands of enzymes, each type controlling one specific chemical reaction—and both macro and trace minerals are integral to their production, and are part of their makeup.

In U.S. Senate document number 264, *published in 1936*, it states, "Our physical well-being is more directly dependent upon the minerals we take into our systems than upon calories or vitamins, or upon the precise proportion of starch, protein or carbohydrates we consume."

We refer to macro and trace elements as if they were two separate forms of minerals. The only difference, and the only reason they have received different classifications, is because macro-minerals are required by the body in large quantities, and trace minerals are required in tiny amounts—sometimes no more than will fit on a pinhead. Because a number of the minerals are called 'trace' it doesn't mean that they are less important. A deficiency of the trace elements can be just as damaging to the health as a deficiency of macro-minerals; likewise, so can too much.

Even a slight deficiency of trace minerals affects enzyme production and this impacts on every physical process including the production of hormones, the regulators of a myriad of other physical functions. Although there is only a very small daily requirement for trace elements, an interruption in the chain of biological amplifications, by the shortfall or absence of a trace element, can be a life or death factor. Current farming methods, diminished trace minerals in soil, food refining, water treatment and the use of pharmaceutical drugs is making it more and more difficult to meet these requirements.

When the digestive system is functioning efficiently, most food-source minerals are, at best, only moderately well absorbed. If there is insufficient digestive acid in the stomach, minerals are poorly assimilated and deficiencies may develop quite rapidly. The aging process and stress both compound digestive weakness and stress itself is a major cause of mineral deficiency as it accelerates the body's use of these elements.

*Ionized Salts*

## ELECTROLYTES

Electrolytes are ionized salts (minerals) found in body fluids and the blood stream. In solution, or dissolved and transformed in water, they can conduct an electric current. When I refer to electrolytes in this book I am talking about trace minerals in solution in water, combined in such a way that they conduct electricity.

Water is the mineral carrier and we absorb them more efficiently when they are dissolved in liquid than we do from foods. Drinking water, once our most important source of minerals, is now an insignificant contributor to our daily intake because it no longer comes directly from lakes or streams; by the time it reaches us it is almost as processed as our foods. Water running over rocks picks up minerals, and the swirling of the water creates vortexes which 'charge' the minerals it carries. The highly processed water running through pipes and into our faucets hasn't got a spark of electrical energy left in it.

The homeopathic process of 'potentising' a substance to create a remedy works on a similar principal to nature's method of creating electrolytes. A homeopathic pharmacist takes a tiny quantity of a ground-up substance, adds it to water and then 'potentises' it by shaking it a measured number of times. This is a parallel to the rushing and tumbling of water that has picked up minerals on its journey over and around rocks.

Once a substance has been through the homeopathic potentizing process, or dissolved and transformed in water, it becomes highly bioavailable and 'alive,' even though it may be so dilute that its presence is almost undetectable.

I supplement my diet, and suggest that my patients supplement theirs, with a true electrolyte formula (Trace-Lyte™ liquid electrolytes). I also recommend fresh organic vegetable juices, sea vegetables, wheatgrass juice, and Volvic™, San Pellagrino™ and other mountain mineral waters as additional trace element sources.

Trace-Lyte™ is the product of a 30 year soil and mineral research project. Each batch is produced using a specific and time-consuming incubation period where the minerals are literally 'grown' together in water. This transmutes them and forms a 'live' electrically charged mineral complex which bypasses the digestive process and absorbs directly into the body, passing through cell walls within minutes of being taken.

## ELECTROLYTES IN ACTION

Electrolytes are essential to the production of enzymes, the function of cells, and in maintaining a normal pH balance in the body and digestive system. Electrolytes also maintain normal fluid balance including osmosis (the cell's internal and external fluid pressure), and blood pressure. But they go one stage beyond this—they bring a special 'aliveness' to the body. To quote Dr. Bernard Jensen, "All cellular structures become alive through electrolytic activity. Life begins with electrolytes. Trace minerals carry the life force in our bodies more than any other substance."

The whole body is a bioelectric organism and the nervous system and brain also operate on electrical energy. Electrolytes are both the 'switch' and the energy source. When electrolytes are depleted, body systems become run down and sluggish, similar to the batteries running a tape recorder. In the machine, like our bodies, while the electrical supply is adequate the system works well, but when run down, it grows weak and slow.

Cells use fatty acids, water, and glucose—and minerals act as positive and negative electrodes producing their voltage potential. Each cell acts as a battery with a different electrical charge on the inside and outside of its wall. The stimulation of a nerve cell sends a wave of depolarization down the nerve fiber, releasing potassium and moving sodium into the cell. As the current passes, the charge difference at the cell wall is re-established as potassium is pumped back inside the cell. This sets up an electromagnetic current which makes up our energy system. It is this subtle energy that produces life.

## *BALANCE*

The balance of minerals (mineral homeostasis) is as important a consideration in health as their availability and assimilation. Minerals can compete with one another for absorption, especially if too much of one is available and not enough of others. For example, too much zinc can unbalance copper and iron levels in the body and large amounts of calcium reduce absorption of magnesium, zinc, phosphorus and manganese.

A similar unbalancing of minerals can occur with excessive intake of single vitamins, either by producing a deficiency or increasing the retention of a particular mineral. David I. Watts, D.C., Ph.D., writes, "a high intake of vitamin C will contribute to copper deficiency as a result of decreasing its absorption or producing a metabolic interference." According to *Biological Trace Mineral Research* (August 1991), the production of free radicals can be stimulated by excess iron, cadmium, or nickel.

A soil mineral imbalance will lead to crops and plants with mineral imbalances, especially if soil mineral levels are unbalanced by artificial fertilizers. When we eat these foods our own mineral levels become unbalanced. Eating refined foods also leads to mineral imbalance. For instance, two teaspoons of white sugar are enough to upset mineral homeostasis and body chemistry. Blood tests before and after ingestion of sugar show that some minerals increase while others decrease—and their relationship to each other is changed. In other words, homeostasis is upset and this opens the door to dis-ease.

Stress, pregnancy, growth, athletic training, sweating, illness, aging and taking medicines all increase normal electrolyte requirements. But most of us aren't even getting the 'normal' levels. Because of our diminishing mineral sources, anything that increases the body's need for them can send health into a tailspin and set the stage for degenerative disease later in life.

The body can tolerate a deficiency of vitamins longer than it can a deficiency of minerals. When you appreciate the significance of this, you'll understand that our need for minerals is as essential as our need for oxygen.

## ✱ ESSENTIAL TRACE ELEMENTS

| | |
|---|---|
| boron | molybdenum |
| chromium | nickel |
| cobalt | selenium |
| copper | silicon |
| fluorine | sulfur |
| iodine | tin |
| iron | vanadium |
| manganese | zinc |

According to Dr. Walter Mertz in a statement outlining the broad definition of an essential trace element, "an element is essential when a deficient intake consistently results in an impairment of a function from optimal to suboptimal and when supplementation with physiological levels of this element, but not of others, prevents or cures this impairment."

## *SETTING THE RECORD STRAIGHT*

The body's delicate mineral balance is easily disrupted. Too many trace minerals in the wrong form and ratios can be as damaging as a mineral deficiency. Some trace mineral supplements are produced from earth or sea beds, and there is a danger when taking these, for an excess of potentially toxic elements to be taken in—including aluminum. Please note that these are *not* what I am referring to when I talk about trace mineral supplements in this book. When minerals are supplied in the right ratio and in a highly absorbable form, as with electrolytes, non-essential trace minerals are chelated *out* of the body.

It is best to add minerals to the soil and eat the plants grown on it, rather than ingesting colloidal minerals in their 'raw' state. Micro-organisms act as an interface at plant roots, ingesting minerals and altering them to a form that plants can use, then plants bio-transmute them to a form *we* can use safely.

The minerals in Trace-Lyte™ are in crystalloid form, a step beyond colloidal minerals. Colloidal minerals are 'inorganic', i.e., they have not been bio-transmuted by plants or by the movement of water. The particles of a colloidal substance consist of very large molecules or large aggregates of molecules that are neither a solution or a suspension, whereas a crystalloid is a substance which forms a true solution and can pass through a semi permeable membrane.

Trace-Lyte™ is time-tested and is the result of almost a century of painstaking research which no other scientist or company has been able to duplicate. The original formula was used in the 1930's and 1940's by a number of doctors working with cancer patients. The FDA was responsible for it's discontinuance in the late 1940's, I dare to add, because it was a valuable addition to their treatment, and they were achieving positive results.

It takes at least a week to make each batch and in the process of its evolution the minerals are literally transmuted in the water. They are combined in a specific way, over a precise period, and during that time something resembling an implosion takes place, creating an energy force that is so powerful it causes the minerals to change form. They are reduced to crystalloid size and then have a unique ability to filter through soft tissue and directly into cells. In this form their potential is extraordinary; they are the very beginning of minerals in action in the body. They are electric. They are alive.

# PIECES OF THE PUZZLE

*'What befalls the earth befalls all the sons of the earth…*
*Man did not weave the web of life, he is merely a strand in it.*
*Whatever he does to the web, he also does unto himself.'*

Chief Seattle, 1853.

## RESEARCHED TO DEATH

Through the centuries we have come to believe that the more complex a medicine or scientific discovery, the more validity it holds. Humbug! While we cling to this dogma we are being led further and further away from health. And in the process we are being researched to death, literally.

Is all the money spent on medical research, science with its artificial products, or medicine with its complex drugs making us any healthier? Even with the camouflage of life expectancy statistics we are, at best, only treading water. And that's looking on the bright side. The statistics for cancer, cardiovascular and other degenerative diseases have been multiplying steadily since the 1950's.

Dr. Alexis Carrel was very much aware of placing too much confidence in mortality statistics as a measure of our health. He said, "All diseases of bacterial origin have decreased in a striking manner…but we still must die in a much larger proportion from degenerative diseases. In spite of the triumphs of medical science, the problem of disease is far from solved."

The population in general must be feeling pretty lousy judging by the amount of advertising and sales of drugs for arthritis, headaches, pain, constipation, insomnia, colds, flu, hemorrhoids, PMS and menopausal problems, etc.

For all its advances, orthodox medicine still doesn't help any one become really healthy.

Medical research and treatment is usually so complex that the most simple and basic elements of staying healthy are overlooked. The body, mind and spirit are so intricate, connected and complex that only a small fraction of *us* is explainable in strict scientific terms. Medicine often misses the intangible bit.

Researchers continue to look down microscopes and investigate toxic chemical compounds to stop symptoms. But disease prevention (beyond the focus on immunization) and optimum health do not receive much attention in a disease-care system whose niche is crisis intervention.

In the long-term this translates to populations who take very little responsibility for their own well-being and, consequently, take no interest in their health until it deteriorates. At this point they choose a form of doctoring that does little more than patch them up until the next crisis.

## THE BIG COVER-UP

There are other powerful problems related to our health and its care. They involve money and politics. Research money is not funneled into anything that cannot be patented (nutrition, minerals, vitamins and herbs) and drug companies have been making it possible for medical universities and research facilities to stay afloat for years. They are certainly not going to 'bite the hand that feeds them' by researching something like the trace minerals chromium and vanadium and their impact on diabetes. You can't patent minerals.

Drug and chemical companies have billions of dollars invested in their products, and huge sums of money are spent on research and development and drug testing. Even when a chemical or drug is proved to be dangerous they are not going to back-down and admit it without a major fight. Because of the financial investment drugs will be kept on the market for as long as is possible, even if proven dangerous. And many a banned drug has found its way to poorer countries despite evidence that it can cause fatalities.

Look at the cigarette manufacturers. With all the evidence that smoking is a leading contributor to the development of cancer—they will never concede. If you talked to an honest member of a big tobacco corporation they may tell you, off the record, "We've been pushing this stuff for years, do you honestly think we are going to turn round now and say, oops sorry, big mistake, we've been killing people. Can you imagine the explosion of law suits?"

Politics is another thorn in the side of health care. In the early 1970's the United States Government invested 30 million dollars on Federal research into nutrition. The resulting report so astounded officials that the copies were yanked before they could reach the public. Why? Because, amongst other things, the *Human Nutrition Report Number 2, Benefits from Human Nutrition Research*, would have made the medical profession look pretty silly. At that time they were denouncing any connection between illness and diet.

One copy did escape however[1], and the summary below lists the conclusions:

A) Major health problems are diet related.

B) The solution to illness can be found in nutrition.

C) The real potential from improved diet is preventative in that it may defer or modify the development of disease states.

This information has been floating about in the political void for 20 years. And until recently, government departments have spent millions more dollars refuting the results of their own research!

If it was known for over two decades that such illnesses as heart disease, many types of cancer and other serious degenerative diseases could be prevented by diet, and the information was hushed up, what else are you not being told about your health—because it may ruin someone's commercial interests?

In Dr. A.F. Beddoe's book *Nourishment Home Grown*, he says, "The USDA put out a handbook, *'Composition of Foods'* in 1950 and then revised it in 1963. In 1975 they put a new cover on it and called it the Handbook of The Nutritional Contents of Foods. To my knowledge there have been no revisions as of 1990. The agricultural chemicals industry, which has been running USDA for decades, probably wouldn't like to see an updated mineral comparison with the 1963 figures."

John Hamaker wrote in a 1977 paper, "The USDA ought to have upgraded its data and included much more trace-element information. Instead, they copied the old 1963 handbook tables and put them out in a fancy new cover in 1975. An honest set of figures on trace elements would show a lot of zeroes on a part per million basis, and damn the chemical agriculture for the monstrous fraud it is."

The disappearance of the trace minerals from our soil, food and water is a prime example of information you have a right to know, but are not being told. If we were healthier and more resistant to disease the medical and pharmaceutical industries would lose unimaginable sums of money. So, let *me* tell you the facts from the beginning...

# BLOWING IN THE WIND

The Native Americans were so in-tune with nature that when the Pilgrims dropped in on them a couple of hundred years ago the land was fertile, contained quantities of minerals and was covered with trees. In the lore and legends of the Iroquois Nation it was said that a squirrel could travel from the Mississippi to the Atlantic Ocean via the trees, never touching the ground. And it was also said that a ground hog could travel the whole width of the U. S. and never touch pure dirt as the top soil was so deep.

At that time land was more or less free for the taking and as forests were cut down and soils were farmed, depleted and gave out (usually within about 3–5 years), people simply kept moving west. But there came a time when this was no longer possible, and that was when the real problems began.

Topsoil, the rich, nutrient laden ground cover, from which trees and vegetation draw their nutrients, averaged 3 feet 200 years ago but is now diminished to about 6 inches.

Over-farming, loss of protective ground cover and trees, open farming, lack of humus and numerous other Agribusiness farming methods have made soils especially vulnerable to erosion by drought, wind and flooding. Many of these factors also contributed to the production of the 'dust bowl' of the Great Plains in the 1930's.

Statistics from the Soil Conservation Service of the U.S. Department of Agriculture show that an estimated 3 billion tons of topsoil disappear into the atmosphere every year. This rate of loss works out to around 7 tons an acre, enough to cover an area the size of Connecticut. Millions more tons end up in the ocean.

We have been whittling away at mother nature for millenniums, but it has only taken us a couple of hundred years to really damage our planet and its ecology.

# PHARM-ING

Professor Justus von Liebig published 'The Organic Chemistry of Agriculture' in 1840[2]. Chemical Companies embraced his tenet of supplying nitrogen, phosphorus and potassium (NPK) to unproductive soils, and began prospering with their artificial fertilizers in the late 19th century[3]. The fertilizers made it possible for farmers to stop moving and stay in the same location—farming the same land indefinitely. This was the inception of today's intensive 'chemical agriculture.'

In the 1920's, the 'all purpose tractor' made its debut, and from that moment on it became easier to rape the land. By the 1930's the food producing acreage was seriously depleted of topsoil and minerals. But that didn't stop the use of farming methods which constantly made withdrawals without deposits[4]. When the winds struck the Great Plains, what little topsoil remained was lost to erosion.

After World War II, the use of unbalanced petrochemical fertilizers was becoming firmly established as normal farming practice. To add to the existing farm chemicals, the chemical arsenal from the war was, metaphorically speaking, beaten into plowshares—chemists hypothesized that what worked to kill bugs on soldiers would probably kill bugs on plants. And it did. Using these 'green revolution' products, including chemical fertilizers, made it possible to obtain high yields of crops, virtually free of insect damage, from what had been progressively unproductive soils.

In the beginning these worked like magic, but by the 1950's and onwards ever more potent chemicals had to be used to produce the same effects. As the next decade dawned, farmers had already pumped around 600 million pounds of chemicals into the soil and the food supply.

It is not coincidental that at this time many chronic disorders like cancer and heart disease began to escalate. Studies over the last few years have shown that the application of chemicals to the soils and plants has a direct relationship to ill-health[5]. Poisonings from foods doused in too much of a particular agricultural chemical have caused death, and growing numbers of people are becoming clinically sensitive to chemicals.

But the greatest and most dangerous consequence of chemical farming is being ignored. The coinciding disappearance of minerals from the soils, their scarcity in the plants grown on them and the deficiencies occurring in people eating them, all results in ill-health, degenerative disease, and an inability to cope with chemicals.

When trace minerals are deficient, the body's defense systems, which prevent chemicals harming cells, tissues and organs, cannot function well. The more chemicals are taken into the body, the more trace minerals are needed. But the chemicals are decreasing the uptake of the minerals. If trace elements are scarce the body holds onto toxic minerals and traces of agricultural chemicals.

Christopher Bird and Peter Thompkins write in *Secrets of the Soil* that, "The activities of enzymes are extremely susceptible to chemicals. The mere presence of chemical additives in food may cause some trace elements to become unavailable. The same applies to chemical fertilizers in the

5

soil, they can cause trace minerals to become unavailable to plants[6]. Enzyme reactions are influenced by a deficit of any functional nutrient."

## AN INTIMATE RELATIONSHIP

Rocks provide the foundation for the creation of topsoil; living organisms slowly ingest them and release the minerals. Soil bacteria thrive when minerals are readily available, and their life-cycle converts these raw elements into a smorgasbord of protoplasm and bioavailable minerals that are taken up by plant roots.

The soil, micro-organisms, and plants have an intimate relationship. The plants rely on the presence of micro-organisms for their nourishment and immunity. The micro-organisms act as an interface, converting the 'raw' minerals in the soil into bioavailable minerals and nutrients for the plants. Like us, plants can build up their own autoimmune system and natural defenses as long as the nutrients are available to them.

Earthworms are also important in nature's scheme of things—their burrowing aerates the land and breaks up hard soils. Their casts act in the creation of living soil because they contain ground minerals and micro-organisms.

Failing to let the land lie fallow occasionally, not plowing back vegetable matter, and allowing soil demineralization results in 'dead' earth with little substance and poor water-holding properties—easy prey to erosion.

Farmers ultimately lost their land when they changed from sustaining the forests, soils and their micro-life, to treating sick plants.

## CATCH-22

In U.S. Senate document number 264, written in 1936, it says, "Sick soils mean sick plants, sick animals and sick people."

Today, after nearly 6 decades of continuing to sicken and deplete the soils, degenerative disease is at an all time high. The U.S. now ranks at the same level as Third World countries with respect to health. And the $750 billion or so that is spent on health care annually in this country has done nothing to prevent Americans ranking 17th in the world in longevity, 19th in general health and 23rd in infant survival according to W.H.O. figures.

On most farms in this country, and around the world,

the soils are not merely sick, they are virtually dead. Earthworms, micro-organisms, trace elements, and humus, are present in only a fraction of their normal levels. Farmers add 46 billion tons of synthetic fertilizer to their crops annually in an attempt to compensate for the useless soil.

The more chemical fertilizers are used, the weaker the plants become, the more insects attack, and the more insecticides have to be used. This creates and sustains the Catch-22 farmers are finding themselves locked into[7].

Mark Anderson noted in his introduction to Dr. Bernard Jensen's book, *Soil and Immunity*, "In modern agriculture, the role of soil reduces to a substance with a convenient texture that holds plants in the vertical position while chemicals are forced up their shaft. Plants sit in the field and receive a chemical enema."

The acidity of artificial fertilizers is very different from the humic acid produced during the natural composting process, and this extreme acidity precipitates the release of quantities of minerals from rocks in the soil. Technically this is beneficial since farm land is so demineralized, but the pH is thrown off balance, soil bacteria are killed and the minerals become unavailable to the plants.

A similar process takes place when acid rain[8] falls on the soil. It also alters the pH and leaches out or locks up soil minerals (even if they are in the soil) so plants and microbes can't consume them. The acidity also kills the soil bacteria which break down and transmute the minerals—and the plants begin to starve. Great forests are dying all over the world because of acid rain. Soil pH is altered, soil bacteria die and minerals are locked up. The trees, like the crops, cannot pick minerals up from the soil and eventually begin to die. In tracts of the Black Forest minerals have been added to the soil and the trees are reviving.

## PLANT PRESCRIPTIONS

Nature is a perfectionist and stops inferior plants from growing. If a plant is weak or damaged, the fungi living with the roots turn on the plant and attack it. The fungus is nature's way of recycling the plants because they don't meet her standards. We have learned to override this wisdom.

Modern farming is a system of producing and medicating sick plants. Farming, like medicine has become a discipline of symptoms—symptoms that are treated with chemicals.

When a farmer finds plants infested with fungus he uses a fungicide. After the application the fungus is usually history, but so too is one of the plant's allies, and the next time the farmer checks his crops they look distinctly sorry for themselves. The prescription calls for more chemicals. The next battalion kills soil bacteria—which cuts off another of the plant's allies and sources of nourishment. Now the plants are sick.

Nature is not amused, so she sends in the heavy squad. Her tough cleanup crew—the insects, or what we call pests. This new symptom is treated with insecticides, to which 440 species of insects in the U.S. are resistant. Insects are really only interested in inferior vegetation, and their ever-increasing numbers should be telling us something.

In her struggle to maintain homeostasis nature, who abhors vacuums and any other imbalances, grows plants (which we call weeds) on unhappy soil. Weeds provide specific nutrients to the soil and their roots can help to aerate and break up hardened and compacted soils. But we apply herbicide.

With the kind of farming we've chosen, weeds are a real problem. Since the mid 1940's, pesticide and herbicide use has risen ten-fold, and at the same time, crop losses have risen two to threefold. We are 'medicating' our sick soils and plants in much the same way that we are medicating ourselves. The result is also degeneration, and a deeper reliance on more and more chemicals just to keep from sliding back too far. It is an interesting parallel with drug addiction...do we need to go 'cold turkey' with the chemicals, bite the bullet for a couple of years, and regenerate our soils with rock dust and compost? I think it has finally reached that point...

# MISSING PIECES

*"It is fair to say that a basic biological difference exists
between natural and synthetic products and substances,
despite their chemical identity and apparent physical similarities.
Extensive laboratory experiments prove this."*

Rudolph Hauschka, D.Sc.

## WHAT YOU SEE IS WHAT YOU GET—
## OR IS IT?

Nutrition charts listing good food sources of specific minerals are a fairy tale. Where and how the food was grown makes the difference between whether the item contains an acceptable level of minerals, or almost none. Most of the time we have little idea of a food's origin.

The Government Accounting Office (GAO) recently re-evaluated Handbook 8, the federal nutrition handbook; known as the yardstick for food/nutrient information. GAO has pronounced it inaccurate and unreliable, stating that in many instances the information is wrong, inconsistent, sloppy and questionable. For example, Handbook 8 credits papaya with 3,000 units of vitamin A, but new tests show there are around 400 units. The report also says that data is often accepted with "little or no supporting information on the testing and quality assurance procedures used to develop the data."

Dr. Firman Bear of the Department of Agricultural Chemistry at Rutgers University tested the mineral levels of 204 samples of vegetables. He was interested in the mineral variations in foods grown at different locations with identical species, growing and harvesting times. The resulting chart *Variations in Mineral Content of Vegetables* is solid evidence that the foods most of us are eating are not what we think they are—and that depleted soil yields depleted crops.

Dr. Firman Bear's Report was published in the *Soil Science Society of America Proceedings* in 1948 and the table in this book is taken from a chart summarizing Dr. Bear's findings, published in the March 1977 issue of *Acres USA*. Decades have elapsed since this report was published, and in the proceeding years our reliance on chemicals has spiraled. We're using superphosphates on our farmlands and greater quantities of diverse chemicals. When you look at the chart, remember that mineral depletion is *far* worse today.

**9**

# THE FIRMAN BEAR REPORT
*Rutgers University*

## VARIATIONS IN MINERAL CONTENT OF VEGETABLES

| | Calcium | Magnesium | Potassium | Sodium | Manganese | Iron | Copper |
|---|---|---|---|---|---|---|---|
| **SNAP BEANS** | | | | | | | |
| Highest | 40.5 | 60 | 99.7 | 8.6 | 60 | 227 | 69 |
| Lowest | 15.5 | 14.8 | 29.1 | 0 | 2 | 10 | 3 |
| **CABBAGE** | | | | | | | |
| Highest | 60 | 43.6 | 148.3 | 20.4 | 13 | 94 | 48 |
| Lowest | 17.5 | 15.6 | 53.7 | 0.8 | 2 | 20 | 0.4 |
| **LETTUCE** | | | | | | | |
| Highest | 71 | 49.3 | 176.5 | 12.2 | 169 | 516 | 60 |
| Lowest | 6 | 13.1 | 53.7 | 0 | 1 | 9 | 3 |
| **TOMATOES** | | | | | | | |
| Highest | 23 | 59.2 | 148.3 | 6.5 | 68 | 1938 | 53 |
| Lowest | 4.5 | 4.5 | 58.8 | 0 | 1 | 1 | 0 |
| **SPINACH** | | | | | | | |
| Highest | 96 | 203.9 | 257.0 | 69.5 | 117 | 1584 | 32 |
| Lowest | 47.5 | 46.9 | 84.6 | 0.8 | 1 | 19 | 0.5 |

Millequivalents per hundred grams / Trace elements ppm

As a result of the way foods are being grown, their nutritional values are altered, but because they still look like grains, lettuces, apples or oranges, we are lulled into a false sense of security.

The contrast in the mineral content of vegetables grown on different soils can be as much as 100 to 1, yet both plants will look almost identical. In addition, the commercially grown 'look alike' fruits, vegetables and grains we buy in the supermarket are likely to contain *elevated* quantities of minerals that are *harmful* to the body.

In his extensive study of the soil and minerals which began in the 1920's, Dr. G.H. Earp-Thomas concluded that when plants are starving, they substitute other elements from the soil. This nugget of information, abridged from his research, illustrates his findings: "If lime or potash is lacking in the soil, or if present in a form nature cannot use, she may substitute, as food for the plant, the minerals magnesium and soda. Or, she may substitute in the place of sulfur, toxic levels of selenium. When the soil is deficient in some particular element and there are no others to substitute, the plants will grow to maturity bearing only a fraction of that mineral." This was confirmed by Dr. Firman Bear with the help of advanced laboratory equipment.

We can relate this to our current health crisis when we understand that essential minerals protect us by preventing plants *and the body* from taking up toxic minerals. Plants grown on properly mineralized soils take up fewer radioactive elements or toxic trace elements than those grown on demineralized soil because they possess the ability to select the appropriate elements from the soil— when there is a full-spectrum to choose from. When the selection of trace and macro minerals is limited, plants then take up quantities of toxic elements. And the same happens in the body.

Superphosphate fertilizers are not only high in the toxic trace element cadmium, but their presence in the soil *stops* the plants taking up the protective trace minerals like zinc.

Cadmium accumulates in the kidneys, arteries and liver where, according to research cited by Dr. Henry Schroeder, it interferes with enzyme systems requiring zinc and, in some cases, can cause high blood pressure which may lead to stroke or heart attack.

How do we accumulate cadmium in the kidneys? Dr. Schroeder provides a clue: "The natural sources (*of cadmium*) are food, water and air. If cadmium can displace zinc in the body, it is likely that food containing more than usual amounts of cadmium and less than usual amounts of zinc might slowly lead to the accumulation of cadmium. (*Sufficient*) zinc would prevent the accumulation of cadmium, a slight deficiency allow it."

David Watts, D.C. Ph.D., writes in the Journal of Orthomolecular Medicine that, "...a deficient intake of an element can allow toxic accumulation of another element. Small amounts of cadmium intake can accumulate to a point of toxicity in the presence of marginal or deficient zinc intake."

Henry Schroeder continues with this insight relating to cadmium induced hypertension in rats, "By injecting a special chelating agent, which binds cadmium more than it does zinc, we can remove some of the cadmium from their kidneys, replacing it with zinc. When we do this, we cure the hypertension overnight. If we continue to feed cadmium, hypertension slowly returns in several months, but can be cured again."

If enough zinc is present in the soil it stops the plants taking cadmium up no matter how high the cadmium in the soil or fertilizer, and in turn the zinc in the plants stops us holding on to cadmium in the body. More often than not though, zinc is unavailable to prevent either the plants or us accumulating toxic cadmium.

## TOO REFINED FOR OUR OWN GOOD

To add insult to injury, in our infinite wisdom, we then intensify the problem and cheat ourselves of even more nutrients by eating refined foods.

Food refining is the process of removing parts of a food. In the course of refining flour the grain is stripped of its bran and germ and virtually all that's left is starch. The body has a hard time processing starch without the trace mineral chromium, fiber, and the other nutrients nature always includes in starchy foods. Pure starch stresses the pancreas and throws the blood sugar into turmoil, and the resulting energy that isn't burned off gets stored as saddle bags and other mainly unappreciated blubber.

By removing the bran we lose trace minerals like silicon and zinc, by removing the germ we lose vital oils, zinc and vitamin E. We are left not just with starch, but also with high levels of the toxic trace element cadmium. As the grains are growing, cadmium accumulates in the starchy part of rice (glutelin) and wheat (gluten)—and that's the part we eat!

Henry Schroeder asks if the refining of foods could remove zinc and not remove cadmium. "The point to be noted is that cadmium is held largely in the endosperm (*the starchy parch)* of the wheat, whereas zinc is held largely in the bran and germ. From these analyses we can conclude that human beings eating white bread, [*white rice*] and white flour products do not get enough zinc to cover their intake of cadmium, or to put it another way, they take in relatively more cadmium than they should for the amount of zinc they get."

In nature's package, zinc found in the germ and bran of grains stops the body from taking up cadmium[9]. In our 'better than nature' product we get a concentration of cadmium without the zinc to prevent its uptake and protect us from the possibility of developing disease.

## FOOD PROCESSING MINERAL LOSS

| Wheat Milling | Sugar Refining |
|---|---|
| manganese 88% | magnesium 99% |
| chromium 87% | zinc 98% |
| magnesium 80% | chromium 93% |
| sodium 78% | manganese 93% |
| potassium 77% | cobalt 88% |
| iron 76% | copper 83% |
| zinc 72% | |
| phosphorus 71% | |
| copper 63% | |
| calcium 60% | |
| molybdenum 60% | |
| cobalt 50% | |

Source: The Trace elements and Man. Henry Schroeder, M.D. (Devin-Adair 1973)

The refining of sugar cane involves removing the fiber and molasses and leaving the starch. Nature put this package together so the minerals in molasses, especially chromium together with B complex vitamins and fiber, assist the body in processing the starch. Fiber slows down the digestion of starch and adds bulk so we can't eat too much. By chucking out the good bits and eating just the starch we're expecting the body to cope with carbohydrate in a concentrated form without the nutrients it needs to accomplish this feat.

A quote from the United States Department of Agriculture serves to illustrate this point: "...the preparation and refinement of food products either entirely eliminates or in part destroys the vital elements in the original material."

# MISSING MINERALS—MUTANT FOODS

When trace elements are missing from the soil and the ground is choked with chemicals, the protein and vitamin content of food changes. A study published in the *Journal of Food Quality* in 1987, showed that produce grown using natural farming methods contains more vitamins and minerals than produce grown commercially using chemical fertilizers. Research carried out at the Universities of Maine and Vermont showed strikingly higher levels of vitamin C, beta carotene, magnesium and calcium in naturally produced vegetables, compared with those grown by commercial, chemical methods.

Researchers have also found that supplying citrus fruit trees with artificial fertilizers lowers the vitamin C content of the fruit, it also lowers the capacity of hybrid corn to produce seeds high in protein. Hybrid grains are generally less easily digested and lower in minerals and protein than ancient grains like spelt, quinoa, amaranth or millet.

In the late 20's and early 30's, Ehrenfried Pfeiffer ran experiments on wheat, growing one batch using chemical fertilizers and the other 'bio-dynamically.' When the wheat was harvested the bio-dynamic grains had 12–18 percent protein compared to 10–11 percent in the chemically produced grain.

Sister Mary Justa of the Holy Rosary College in Buffalo, New York, conducted many experiments with foods. Her work with electronmicroscopy illustrated the difference between natural and synthetic vitamin C. At one point in her research she grew corn, some using natural farming methods, and some using the commercial chemical methods. She found that the protein quality and level was higher in the organically produced corn.

Specific trace minerals enable plants to produce protein. Henry Shroeder wrote in his book *The Trace Elements and Man*, "Molybdenum is required by legumes for it is an essential element for the growth of nitrogen-fixing bacteria on their roots. These bacteria convert atmospheric nitrogen to soluble nitrates which are absorbed by the plants to synthesize proteins."

In *Bread from Stones*, Dr. Julius Hensel maintains that a lack of trace minerals actually changes the molecular structure of protein. A. F. Beddoe writes, "The protein content of food is high or low in about the same proportion as the minerals. This is because just about all the minerals are used in the amino acid enzymes which in turn are catalysts helping to make all the protein compounds."

All of this has unfortunate implications for vegans who rely solely on beans, grains and vegetables for their protein, and for people who are vegetarian and rarely eat eggs, yogurt or fish. I see many vegans and vegetarians who come to me because they just don't feel well. Specific supplementation including sources of amino acids with electrolytes has brought them back to feeling healthy again.

## SIDE EFFECTS OF SIDE EFFECTS

As if all you've just read isn't bad enough, we add insult to injury by taking drugs like corticosteroids and diuretics which have a catastrophic effect on mineral and vitamin levels in the body. Physicians and drug manufacturers concern themselves with the conservation of potassium during treatment with diuretics—as if it were the only mineral affected.

A snoop at the more technical pharmaceutical books shows potassium to be only one of many minerals lost. Macro-minerals including magnesium are excreted in significant amounts, especially with thiazide diuretics, so too are vital trace elements including zinc. This loss, in the long-term (diuretics are generally taken long-term), increases the chance of kidney and cardiovascular damage, precisely the organs that are malfunctioning when the physician prescribes diuretics for high blood pressure or edema (if the problem is not a side effect of HRT or other medications).

But the story doesn't end there, because a loss of magnesium prevents the body from utilizing potassium properly, another indicator of how important mineral balance is in the body. Impaired glucose tolerance is a feature of long-term diuretic treatment and it would most likely not occur if potassium and chromium levels were undisturbed. By rebalancing both potassium and chromium this side effect can be reversed.

It is a standard medical joke that taking vitamin supplements results in expensive urine. It's no joke that taking diuretics and many other drugs results in priceless urine! The cost is not counted in dollars but in lost trace minerals and ultimately the patient's health.

How many people do you know who take diuretics or other medicines and have concurrently been prescribed a multiple vitamin or mineral supplement? It occurred to me a long time ago, that if we are going to mess around with chemicals, or manufacture them as drugs and put them in our bodies on a regular basis, then we should be sure about what each one is capable of doing to the body's nutritional status, and what the combination of several of them can do. Otherwise, how on earth can we expect to regain real health?

Drugs may deplete minerals by increasing their excretion, by interfering with mineral balance, or by antagonizing synergistic factors. An example of this is explained by David Watts, D.C., Ph.D., in a paper published in the *Journal of Orthomolecular Medicine*, "...antacids, laxatives, anti-convulsants, corticosteroids and antibacterial agents

| MEDICATION | MINERALS AFFECTED |
|---|---|
| Amphotericin Antifungals | potassium |
| Antidiabetics | iodine |
| Antacids | calcium, magnesium and phosphorus |
| Anticonvulsants | calcium and magnesium |
| Antiparkinsonism Medicines | potassium |
| Antithyroid Medicines | calcium and phosphorus |
| Aspirin | calcium, magnesium, phosphorus and potassium |
| Bronchiodilators | potassium |
| Corticosteroids, Prednisone and Prednisolone | calcium (excessive loss from bone), phosphorus, potassium and zinc |
| Digitalis | magnesium, potassium and zinc (Digitalis toxicity is provoked by potassium deficiency) |
| Diuretics | calcium, chromium, iodine, magnesium, phosphorus, potassium, sodium and zinc (most trace elements) |
| Cholesterol Medications | iron |
| Penicillins | magnesium and potassium |
| Sulfonamide | calcium and magnesium |
| Tetracyclines | calcium, iron, magnesium, potassium and zinc |
| Thyroxine | calcium and magnesium |
| Ulcer Medications | iron and potassium |

are known to produce a deficiency of calcium and vitamin D. They exert a chelating action upon calcium and antagonize the metabolic effects of vitamin D. Prolonged use can lead to rickets, osteomalacia and other calcium deficiency disorders."

Because the trace elements are just that—trace, they are rarely considered in drug studies. The data on iatrogenic (illness induced by a doctor's treatment or drugs) trace-element loss doesn't really exist, with the exception of zinc. However, when one mineral becomes deficient in the body, the balance or homeostasis of the other minerals is disturbed.

There is another ironic twist to the problem of drug induced vitamin and mineral deficiencies: long term therapy often causes depletion of exactly the nutrients the body needs to recover from the problem that is being treated. You remember that diuretics deplete the nutrients necessary for the kidneys and cardiovascular system; these same nutrients, especially potassium and magnesium, help to normalize blood pressure.

Antibiotics affect mineral absorption, and minerals play an integral role in the health of, yes—you guessed it, the immune system. According to Henry Schroeder M.D., antibiotics block the absorption of macro and trace minerals. This includes zinc, which is well known for its importance in the production and health of T-Lymphocytes, the body's search-and-destroy cells in the immune system.

With this understanding, it makes sense why the lack of zinc, and the reduction of healthy intestinal bacteria leads, with repeated doses of antibiotics, to a weakening of the immune system and increased vulnerability to opportunistic infections and yeast infections like candida. Additionally, the toxins candida produces inhibit the function of the T-Lymphocytes.

## THE UNSUSPECTED THIEF

Eating refined sugar causes the body to use and lose more trace minerals than people realize. And the result is ill health. Some foods are 'anti-nutrient,' in other words, they contribute nothing to the body in terms of nutrients, and while in the body, wreak havoc with *existing* nutrients. White refined sugar is little more than starch in a very concentrated form. When this arrives in the body, the pancreas and liver are forced to draw on vitamin and trace mineral reserves, especially chromium, in order to process it. At the same time, pH is thrown off, tipping the balance towards acid. In order to maintain balance or homeostasis, the body pulls more minerals from its reserves. At every stage of sugar metabolism minerals are withdrawn. White flour has a similar effect as it is almost entirely starch.

Few people are aware of junk food's enormous drain on trace mineral reserves, or even of the amount of sugar they are consuming. A great deal of sugar is 'hidden' in foods like bread, condiments, pre-prepared meals, meats and more, and a consistent intake drains the body's mineral reserves. Because minerals are so scarce in soils and food, even if sugar is cut out of the diet, the trace mineral levels may take time to go back to 'normal' if at all.

The sugar kids are eating in fast foods, junk foods, candies and sodas is setting them up for hyperactivity, ADD, learning disorders, depression, aggressive behavior, and degenerative diseases like cancer, heart disease, arthritis and osteoporosis later in life.

Violence and aggressive behavior in schools has reached astronomical heights. The lucky kids eat a decent breakfast, most eat sugar drenched cereals, or none at all, and many eat candies or junk on the way to school. By the time they reach the classroom many are off the wall. It doesn't take much to put them over the edge. Low blood sugar can result in temper outbursts and violence. We can go on policing schools and building jails till the country is bankrupt, but what we really need to investigate is an answer to the social issues, and stopping the refined sugar while putting the minerals back in our diets.

## HAVE WE LOST OUR MARBLES?

Maybe we have. Trace minerals are missing from the soils. The crops we grow for food are starving, chemical ridden and sick. Foods are stored, refined and cooked (and the water is thrown away). We take drugs that radically deplete minerals and live at a time when our need for minerals is the highest it has ever been.

The typical American menu and processed water supplies contribute marginal amounts of trace elements, and carry toxic minerals like lead, cadmium and mercury along with toxic chemicals. The resulting mineral deficiencies and toxic overload can short circuit the normal workings of the brain and cause subclinical symptoms of reduced mental ability. The brain must have trace minerals to burn blood glucose properly. Seriously low levels of electrolytes can produce severe mental confusion.

Scientists are aware that trace element deficiencies can affect both mind and body; in fact, every physical and mental process. Electrolytes are the sparks which fire and fuel the neurotransmitters that enable us to think and process information. They are vitally important for good brain function and sanity. They prevent us from losing our marbles!

# PUTTING THE PIECES
# BACK TOGETHER

*"Once Americans begin eating food grown on soil containing
all the essential elements, disease will practically vanish…"*

❦

Jonathan Forman, M.D.

## *Adding Life To Your Years
## And Years to Your Life…*

Most people eat so atrociously it's a wonder they are still standing up. And I think we have reached the point where even if people ate organic foods exclusively, there would still be no guarantee that all trace mineral requirements would be met.

Organically grown crops have the advantage over their commercial counterparts. Their chemical content is much lower and they are grown without artificial fertilizers, the acidity of which can prevent the uptake of minerals. But soils are running out of minerals and organic crops can only contain the elements that are present in the earth; compost only recycles what is already there.

Until farmers re-mineralize their land we will have to supplement our minerals, eat the best food we can find, and drink unpolluted spring or mineral water—that is if we want to be healthy. All the medicines or vitamin supplements in the world are not going to do it for us. Neither are the doctors or the government. We are generally kept in the dark and about health and nutrition, especially if it relates to unprofitable information about food's ability to keep us healthy, or anything which may threaten the 'disease' industry's revenue.

We are also kept ignorant about farming practices that contribute to the upsurge in degenerative disease.

Joel D. Wallach, N.D., states that, "Since the early 1900's veterinarians have known all about deficient soils…Since the 1950's, veterinarians have known that the risk of cancer, heart disease, cataracts, liver cirrhosis, muscular dystrophy and other chronic disorders could be reduced to almost zero simply by adding selenium to animal diets. In fact, muscular dystrophy made the live-stock industry unprofitable in the high rainfall areas of the U.S. prior to the 1950's because of selenium deficiency. It is interesting, but very sad, that farmers could then grow crops for human food on these deficient soils by adding just nitrogen, potassium and phosphorus (NPK). When the mineral depleted soil couldn't support animal life they grew human food."

## WATER, THE FORGOTTEN LIFE-FORCE

River water has kinetic energy beyond our normal perception. It is one of the many natural forces in nature with which we have lost touch. We can become aware of them through observation. For instance, birds navigate using electromagnetic forces from the earth, and there are forces in water which fish use. We tend to think of water as a lifeless substance, only animated by movement, but in Olof Alexandersson's book *Living Water* he talks about how Victor Schauberger, an Austrian forest warden, observed water's hidden energies.

Some of Shauberger's earliest observations involved mountain trout. On one occasion he was sitting by a waterfall watching the fish in the moonlight when a large trout began a strange almost dance-like movement. He noticed it next in the jet of the waterfall, briefly in a conically-shaped stream of water and then floating motionlessly upwards till it reached the top and swam off against the current. Schauberger noted that energy can build up in streams, flowing in the opposite direction to the water.

The movement of water and the minerals in it can interact to produce extremely powerful energy—in an electrical form. In an other instance Schauberger was sitting by a rock pool formed in a stream, and noticed that there were some football-sized, egg-shaped stones moving in the bottom, in fact, 'dancing' in a similar way to the trout, and then 'floating' on the surface of the water.

The energies in water can be extremely powerful and, in nature, can be involved in the conversion or transmutation of minerals to another form.

## MINERALS AND WATER

Once inside a living cell, water takes on different properties with immense amounts of 'potential energy' that can be triggered to perform work. As this energy is triggered a cell moves, conducts electric impulses, transports solutes and synthesizes chemicals.

The latest research shows that minerals are absorbed most effectively by the body when they are transmuted in a liquid medium.

Martin Fox, Ph.D., author of *Healthy Water for a Longer Life* says that the chemical form of certain elements in water is different from those found in food. He also points out that there are agents in numerous foods that prevent the active assimilation of some minerals so, even if their content in the food is high, their absorption may come down to very little. Medical researchers agree that you absorb toxic minerals far more efficiently when they are dissolved in water and so, by the same mechanism, are the trace minerals we need.

Minerals dissolved in spring or artesian water are considerably more bioavailable than those present in food because they are already in solution[10]. Water is the ultimate mineral solvent and breaks them down into their free ionic state where they become electrically charged—electrolytes. In this form they are easily absorbed through the walls of the gastrointestinal tract, and then by the cells[11].

The Vilcabambas of Ecuador and the Hunzas of Pakistan are both known for their exceptional health and longevity. These people drink, and grow their crops in, mineral water flowing from the glacial backdrop to their villages. The mountain streams cascade through mineral beds and swirl over and around rocks creating vortexes (which look like underwater tornadoes). This action generates an electrical charge in the minerals taken up by the water, converting them to another form and making them exceptionally bioavailable to both plants and the body.

In *Secrets of the Soil*, Peter Thompkins and Christopher Bird elaborate on the Hunza water and the reason why it contributes to longevity. They also explain that the minerals in naturally flowing water carry an electrical charge (usually negative). They too, propose that it is this which enables the body to absorb the minerals even though they have not been processed organically by plants or animals.

## 'ORGANIC' AND INORGANIC MINERALS

Natural iron supplements such as iron peptonate will also absorb into the cells particularly if the body's electrolyte balance is correct. But many inorganic supplements and even food 'additive' supplements do not. A recent paper in the *European Heart Journal* (1992) pointed out that the body does not absorb inorganic minerals[12], and as a result they can become deposited in the arteries and joints, leading to degenerative disorders like heart disease and arthritis.

The scare over iron's link with heart disease is explainable from this standpoint as inorganic iron taken into the body stays out of solution (out of homeostasis) in the blood.

Iron sulfate must, by law, be added to all foods where the iron and other minerals have been removed by processing. Isolated or random minerals added to food after the natural ones have been processed out causes an imbalance, one that is echoed in the body.

Hundreds of supermarket foods like breakfast cereals, cookies, pie crusts, brownie mixes, breads, burger buns, ad nauseam, ad infinitum, contain unbalanced and unabsorbable iron sulfate. In this form it has the potential to become a component of the atherosclerotic plaque which builds up on damaged arterial walls.

The *form* of iron is only part of the problem—it is primarily the body's inability to absorb it, and this is compounded by a lack of trace elements. Without electrolytes, iron can float about like a 'free-radical' in the bloodstream. Electrolytes help to chelate toxic and inorganic minerals out of the body before they cause damage, and they assist in the absorption of organic minerals into the cells where the body can use them.

## TO DISTILL OR NOT TO DISTILL, THAT IS THE QUESTION

Water, like most of our food, is processed before it reaches us, and unless it has come directly from a well or spring, it will probably have received a good dose of chemicals too. Many people are aware of this problem and use distillers, reverse osmosis, or ion exchange units to produce pure water, free of chemicals *and* minerals.

There are a number of authors who claim that all but distilled or pure $H_2O$ leads to health problems. I believe that beyond the important question of chemicals and pollutants—they have missed the point. Their argument is based on the knowledge that inorganic minerals cannot be metabolized well by the body and are therefore deposited everywhere but the right places. With regard to tap water, and up to this point, I agree but nature's power to convert minerals into a bioavailable form in naturally flowing water has been discounted. Avoiding unpolluted spring and mountain waters deprives the body of one of the most bioavailable sources of minerals.

Until the technological age people drank spring and well water, with the occasional inclusion of rainwater (which is naturally distilled). Scientists now agree that materials dissolved in water become exceptionally bioavailable, easy for the body to absorb and use.

This is true for toxic, as well as macro and trace minerals, but nature has a way of protecting us as long as we keep our meddling selves out of the picture. Dr. John Sorensen, a medical chemist and leading authority on mineral metabolism found that the body's use of nonessential or toxic minerals is affected by the amount of essential minerals present in water. If the essential minerals are present, little or none of the toxic element is absorbed. For example, if there is high calcium and magnesium to lead, the body will select calcium and magnesium and excrete the lead—and vice-versa.

Homeopathic remedies are made by grinding substances into powder, dissolving them in water and shaking to 'dilute' them. This gives the material an extraordinary bioavailability, although the actual concentration of materials may be so minute that, technically, they should have no significant effect on the body. The same is true of minerals dissolved in spring water (electrolytes) even though their concentrations may be so low as to be dismissed as useless.

Foods appear to have higher concentrations of minerals, but these are not always taken up and used efficiently by the body. There are chemicals in some foods which limit their uptake, and most of the foods are grown on demineralized soils. Add to this the processing, chemical disinfecting and, in some states, treatment with aluminum (which disrupts trace-mineral absorption) of our water supplies and it becomes painfully obvious that we are missing one of the most important sources of trace minerals, because we figure it doesn't count!

## DEGENERATION/REGENERATION

Minor aches, pains, anxiety, sleeplessness and the manifestations of degenerative disease are simply the body's way of telling us that a problem is brewing. They are not a sign that we are deficient in pharmaceutical substances like aspirin, corticosteroids, barbiturates or antacids.

Take arthritis for example; the initial symptoms of this degenerative disease are usually ignored until the discomfort has to be treated with pain killers and anti-inflammatory drugs. If these were only a temporary part of the treatment it would make far more sense than the reality—that they *are* the treatment.

Arthritis is one of the long-term effects of mineral imbalance and free radical damage. It is the manifestation of an acid condition in the body, and develops because minerals are out of homeostasis, out of solution in the blood stream, and depositing in joints. With dietary changes[13] and supplementation with liquid electrolytes, calcium, magnesium and other forms of trace minerals, the pain and degenerative processes can usually be arrested, even the pain of advanced arthritis and its progressive degeneration can eventually be checked and eased.

My research and work with patients has been enough to convince me that all physiological causes of disease can be related to mineral and electrolyte imbalance somewhere in their inception and development.

A 38 year old woman came to see me because she had progressive arthritis, to the point where she could hardly move on occasions. She was taking pain killers too numerous to mention, and repeated courses of antibiotics for an assortment of tenacious infections. She had been the recipient of tranquilizers but had taken herself off them as she couldn't stand the side effects, she was depressed but 'hyper' and couldn't sleep and was convinced she was becoming senile as she couldn't remember recent events, people's names, or phone numbers. Her libido was nonexistent, and headaches were something she had learned to live with. That was on top of constipation alternating with diarrhea, incessant stomach problems and a love-affair with antacids.

She told me she had given up hope of ever feeling better and that medicines seemed to work less and less the more she took. As I continued to question her and perform a number of naturopathic diagnostic techniques, her problems began to emerge as a major mineral imbalance. I was puzzled because, although the diarrhea would exacerbate electrolyte imbalance, her diet wasn't as 'off-base' as I had thought, and she wasn't a dedicated drinker, smoker or imbiber of outrageous quantities of coffee.

It wasn't until I asked whether there were any particular foods she craved that we got our answer. She thought about it and then said, "Does salt count?" When I found out how much salt we were talking about I assured her it did. Four ounces of salt a day or more, half a juice glass of salt...every day.

When I had recovered my composure I asked her how she managed to perform such a feat. The story is intriguing. She remembered that when she was 14 she started to crave salt (this coincides with puberty, a major growth and development period when the body's need for vitamins and minerals is intense) and she has been eating it ever since. She even mentioned that sometimes her cravings could be so strong that she would go out and eat chunks off the salt licks in the stables at her parent's home, and that as she got older she started craving beer with salt *in it...*

My conclusion was that the salt craving was nothing more than her body trying to tell her it was desperately in need of minerals. She just supplied it with the wrong ones, and as a result, had become the paragon of electrolyte imbalance and homeostatic disruption. She thought for a while and then said, "All these years with the pain and the drugs, and you think the problem is a lack of minerals and too much salt?"

A week or so later I received a message on the answering machine at my practice. "Hello this is Jodi. I've been using the juicer, taking the minerals and doing all the things you suggested. My bowels are much better, my digestion is better too, and my constant headache has gone. You're not going to believe this, but I haven't craved salt for two days and if I do eat any it makes me feel sick, how did I ever eat so much? I also felt a little less stiff when I got out of bed this morning and am sleeping better than I can remember." All that pain, and simply because her body was trying to tell her that it needed minerals.

Jodi is an extreme case, but as I noted earlier, mineral imbalance can be at the bottom of all physiological disease. Conditions like PMS are linked to an electrolyte imbalance because trace minerals are needed for normal glandular health and hormone production.

The stress reactions triggered during PMS episodes deplete minerals. In men, one of the main physical causes of impotence is atherosclerosis of the penile arteries which restricts blood flow, and this can also be traced back to mineral imbalance. Over time, electrolyte supplementation can 'break down' mineral deposits wherever they are in the body.

Simple problems like muscle cramps and spasms have their origin in calcium, magnesium and electrolyte deficiency, and even serious age-related disorders like deterioration of brain tissue, or senility, are linked to electrolyte imbalance. I could go on and on...cataracts, fungal infections, insomnia, glaucoma, allergies, digestive upsets, anorexia, immune disorders, fatigue, eczema, psoriasis. In order for the body to effect rebalancing and regeneration it needs minerals and electrolytes in conjunction with any other treatment. They are the catalyst for healing.

In the areas of the world where quality of life and longevity are legendary, it traces back to the high mineral content of the soil, plants and water.

It is a scientific fact that the human body has the potential to stay healthy for approximately 120 years. In our 'advanced' culture where we rely on chemical and drug companies for our existence, too many people spend the second half of their lives suffering from the symptoms of degenerative disease, losing their independence, and handing over life savings for drugs, hospitalization and 'new parts.' According to Dr. Walter Bortz of Stanford University "We live too short and die too long."

## GREAT-GRANNY'S SECRET

Our great-grandparent's nutritional staples were the dietitian's nightmare; whole raw milk, butter, cream and cheese. But they did not die by the hundreds of thousands from cancer and heart disease.

Even so, this is not an authorization to guzzle a gallon of *Ben and Jerry's* ice cream or reinstate the lard—it simply illustrates the point that something has happened between then and now that has made us more vulnerable to degenerative disease. Yes, stress, chemical and pollution levels were lower, and yes, the majority of people spent more time engaged in physical labor, but the fact remains—despite the high intake of dietary fats in those days, cancer and heart disease were rare.

Food was not processed, pasteurized or cooked as much as it is today, and white flour and white sugar were primarily the indulgence of the affluent. The 'pure' white products were too expensive to be part of the basic diet of most people. Foods were usually left whole and grown without chemical fertilizers. Water supplies came from springs, streams or wells and contained minerals that the body was able to use. Dr. Jeffrey Bland notes that our ancestors obtained between one-fourth to one-third of their minerals from the water they drank. In other words, our predecessors had a far greater mineral intake than we do. They acquired the electrolytes needed to keep the cells healthy, activate the body's enzymes, glands, defenses and antioxidant enzymes and process fats properly.

Since the 1950's the death-rates from cancer and degenerative disease have skyrocketed. The establishment of pervasive chemical farming, and the availability of cheap refined and processed foods were the genesis of today's overwhelming occurrence of cardiovascular disease and cancer. They also mark the point where the already diminishing trace minerals almost vanished from our diets.

# ELECTRIC MINERALS

In my practice I recommend the trace element supplement Trace-Lyte™. This formula is a 'live,' electrically charged mineral complex which, because it is crystalloid (a stage beyond colloidal), bypasses the digestive process and absorbs into the body, passing through cell walls within minutes of being taken. In this form, the preventive and rejuvenative potential of trace elements becomes infinite.

One of the first 'side effects' of restoring trace minerals to the diet, in electrolyte form, is improved digestion. This influences every aspect of the body's function. The delicate acid/alkaline (pH) balance of the digestive system is kept stable by minerals, resulting in the production of digestive acids, enzymes and good fat, protein and carbohydrate metabolism. Electrolytes aid in the conversion of proteins to their component amino acids. The level of nutrients taken up, and manufactured by the body is also increased and elimination is improved.

The production of digestive enzymes is impaired if minerals aren't available. The ptyalin of saliva is essential for oxidizing starch to sugar and functions under alkaline conditions (created by specific minerals). A deficiency of ptyalin results in incomplete digestion of starches and intestinal fermentation. This limits the production of lactic acid, alters pH, and creates the kind of environment that putrefactive bacteria and candida love, and acidophilus bacteria struggle to survive in.

A healthy colon is acid, but a population explosion of putrefactive bacteria produces an alkaline environment which, apart from being unfriendly to acidophilus, is irritating and destructive to the delicate lining of the digestive system.

As putrefaction takes place, food decomposing bacteria produce toxins that are absorbed into the circulation. Think of it like a woodburning stove using 'high tar' wood. The smoke in the pipes leaves deposits and decreases its efficiency. In our bodies, circulating toxins reduce the body's efficiency and add to the risk of developing arteriosclerosis, kidney, gall bladder and liver disease.

An equally important result of improved digestion and stable pH levels is the creation of a favorable environment for the growth and proliferation of acidophilus bacteria. The body manufactures vitamins, acidophilus and enzymes and breaks down fats and proteins when electrolytes are present. Acidophilus bacteria will not grow in the intestines unless the body's pH is virtually perfect—and only minerals can accomplish this.

# CHOLESTEROL AND CARDIOVASCULAR DISEASE

*"Progress can only be made by ideas which are very different from those accepted at the moment."*

&#9866;

Hans Selye

## *WHAT THE EXPERTS MISSED...*

Whenever we fool with nature we lose. And we are at it again. Because we have messed around with our soils and food supply so completely we are suffering from a plague of degenerative diseases, and the 'experts' are telling us that we can prevent some of them by taking our food apart or swallowing aspirin(s) every day[14].

Left to her own devices nature produces plant and animal life in a balanced state, but we have decided that we can do better, fix all her mistakes, and live happily ever after. To this end, the role of health villain has been ascribed to saturated fats. The margarine producers are rubbing their hands together with glee, sales of skimmed milk have delighted the dairy industry and packaged food manufacturers have jumped on the no-cholesterol, 'lite' bandwagon.

Now, before all of you who believe implicitly in the no-fat theory come after me with a noose, I'd like to point something out. There is a definite link between some forms of fat, elevated cholesterol, and cardiovascular disease but, as Dr. Lendon Smith says, "The link is not as clear cut as we would like. This ambiguity is displayed when the statistics point out that one-half of the people who die of a heart attack have a 'normal' cholesterol level (200 or less)."

The positive aspect of this current dietary trend is that people are eating low-fat and no-fat foods and absorbing fewer pesticides, dioxins and other malignant chemicals that get stored in the fat of animals and animal products. They are also decreasing their intake of solvent-extracted, hydrogenated, fractionated, heated, microwaved, fried, pasteurized and homogenized fats—all of which give the liver an extremely rough time and increase the risk of both cancer and heart disease. But note—these are all fats that have undergone some kind of human interference.

My research and work with patients has convinced me that it is the lack of trace elements which enable the body

**29**

to use fats, and the prevalence of palm and coconut oils, and the adulterated fats, foods and drinks, alcohol, and stress that are the major contributors to the increase in degenerative disease.

Which leads me to the negative aspects—people have been so bombarded with information about how bad fat is, they are becoming terrified of it. To the point of avoiding any organically raised, unprocessed and natural food products containing fats. This results in an extremely lopsided diet, decreases the intake of yet more trace elements, and downplays other important and more basic reasons for ill health. As Dr. Lendon Smith says, "Remember, we need cholesterol as the basic structure to form cortisone and male and female hormones, also as one of the helpers to combat free radicals." Fat also helps in the assimilation of fat soluble vitamins like A, D and E and keeps the skin moist and healthy.

## WHICH CAME FIRST—THE CHICKEN OR THE EGG SUBSTITUTE?

Eggs have received a real beating lately. Please pardon the pun. I can forgive you if you think they have been put on this earth to slowly and insidiously kill us all off. After all, that's what the medical profession and the media would have us believe.

Beyond the hype there is actually not one shred of evidence to prove that eggs cause cholesterol problems. How is it that our great grandparents, and their great grandparents survived for years on all the eggs they ate? How is it that the French, who indulge in quantities of omelets and quiches beyond our wildest imagination, do not expire in staggering numbers from egg-induced cardiovascular disease? The answer is simple. When eggs are eggs, not mass-produced apologies for eggs, and trace minerals, *the key to this mystery*, are in the foods and water then fats are processed properly in the body. They don't hang around in arteries waiting to do damage.

Keeping hens in tiny boxes, feeding them unmentionable garbage, injecting them with antibiotics and hormones, denying them sunlight and fresh air, and making their lives utter misery results in their laying eggs that are higher in cholesterol than normal. Maybe the hens are getting their revenge! Eggs from naturally reared chickens are lower in cholesterol and arrive balanced by nature with plenty of lecithin; a natural fat emulsifier. When eggs are lightly cooked, the lecithin remains—and nature takes care of the fat.

Part of the medical answer to the cholesterol problem is to advise people to stop people eating eggs and consume that gross stuff in packets. Egg substitutes are little more than chemicals. Eating egg whites without the yolk is not the answer either, it's like throwing the baby out with the bath water. The *main nutrients* are found in the egg yolk.

In response to the food industry's brilliantly worded and slanted advertising people are petrified that butter is going to kill them too, and are buying mountains of margarine, other 'no-cholesterol' spreads, and highly purified vegetable oils (pure fat, as nutritionally useless as white sugar) in the belief that they are choosing healthier products. They are not.

Processing unsaturated oils and converting them into margarine (primarily by hydrogenation) enables them to promote the development of arteriosclerosis even more efficiently than naturally saturated fats like raw butter. Hydrogenation exposes fats to temperatures sometimes exceeding 1000°F which twists the molecules changing their shape from a 'cis' to a 'trans' form. Natural fat or oil (lipid) molecules 'fit" into cell membranes and aid in their function, but heat twisted molecules (trans fatty acids) don't fit, can get locked into the structure and prevent it from functioning properly—at the same time as preventing the cis fatty acids from doing their job. Margarines can be as much as 35 percent trans fatty acids.

For heaven's sake, if we believe the medical profession's interpretation of nutrition and its relation to health—the human race should have been extinct 2000 years ago. Our health problems are not caused by foods we have been eating for centuries. They are caused by a lack of trace minerals which hinders our ability to digest and utilize food components, and our persistent meddling with nature and our foods.

## HIDDEN FACTORS

Funny, but no one is telling us that smoking, eating white sugar, white flour, processed oils, a predominantly cooked diet, and drinking coffee and alcohol can contribute to high cholesterol and triglycerides. But then that would be treading on some rather large toes.

John Yudkin comments that sugar consumption parallels the rise in coronary artery disease (CHD). In 1972 researchers from the University of Hawaii fed 80 pigs high refined sugar diets, 68 of the pigs developed heart disease. I believe this is a result of refined sugar leaching trace minerals from the body. It is also my suspicion that the astronomical consumption of sugar in this and other countries is responsible for much of our CHD. This is compounded by the lack of usable fatty acids in our diets, not only because the fats consumed are mostly altered and 'foreign' to the body, but also because the trace minerals are not available to enable the body to use them.

Anyone who has seen the film, *Lorenzo's Oil* will have an idea of my theory. If the body does not receive what it needs from the diet, and it has the capability of manufacturing whatever it needs on its own, it may get too enthusiastic and manufacture an excess. In a recent article, Dr. Lendon Smith points out that "Cholesterol can be reduced by the ingestion of an electrolyte solution and the consistent use of essential fatty acids."

## TAKING OUR FOOD APART

We are all sold on the idea that taking our food apart, especially removing its fat, is a healthy option. Have we really thought about the logistics of this? Henry Schroeder reveals the scientific reasons for keeping foods whole in his book, *The Trace Elements and Man*. He explains that fractionated foods are undesirable because the mineral balance is disturbed. This is even more serious in commercially produced foods because the mineral balance is already off the wall.

Dr. Schroeder gives some interesting examples of what happens to food's mineral balance when it is 'taken apart.' For instance, skimmed milk has lost virtually all the chromium and manganese, both of which are needed to metabolize its two main ingredients; sugar and protein. He goes on to explain that the only way to obtain a balance of minerals from fractionated foods is to combine them. Mixing corn oil with skimmed milk creates a revolting, but balanced product with respect to trace minerals. However, neither one alone is adequate because they are both 'orphans.'

If you stop and think about it we have witnessed all kinds of silly ideas about nutrition over the years. Now the runaway train is thundering even faster down the tracks. We are buying into another dietary monster.

In the 1960's Dr. Schroeder warned that if trace elements, especially chromium, are missing from the diet, fats will not be broken down or metabolized properly. Trace elements encourage good liver function, improve digestion and metabolism of fat, and ensure that minerals are kept in solution in the blood so they won't deposit on damaged artery walls.

# ESKIMOS, CLOGGED ARTERIES AND ENZYMES

A highly mineralized diet keeps arteries clean by triggering enzyme production. Enzymes enable the body to use fats. Electrolytes enable the body to make enzymes. The medical profession has either missed this or is completely ignoring it.

In *Enzyme Nutrition,* Dr. Howell states that, "The underlying cause of atherosclerosis reverts to the mal-digestion of fat in the digestive tract and the absorption of defective fatty products." He also points out that cooking or heating food kills the enzymes which assist in the digestion of fats, proteins and carbohydrates. This puts the onus on the body to produce enzymes. The standard American diet forces us to borrow from the body's supply of trace minerals and enzymes every time we eat. Few of us have a reserve of trace minerals to draw on, and fats are not broken down properly—they are clogging up our arteries instead. This is the key factor in the current cholesterol problem.

Dr. Howell notes that Eskimos sometimes eat several pounds of fat a day. If you believe what we are being told, they should have extraordinary cholesterol problems, but visiting medical teams have found their arteries to be clear—unless they have corrupted themselves with our 'foods' for a number of years. Eskimos eat raw meat, they let it sit for a while so its natural enzymes 'pre-digest' it. They also eat a great deal of sea food—and vast quantities of our land minerals have found their way into the sea. Eskimos can handle fats because they eat enzymes *and* minerals at every meal.

We are told to avoid fat. If fat alone really was the problem, then Eskimos should have our health statistics, and we should be aspiring to theirs. But that's not the case. The Eskimos are fine and it is us who are fighting an uphill battle with heart disease. Natural fats accompanied by trace minerals are utilized and excreted, they do not contribute to atherosclerotic plaque. The adult body must have fatty acids to remain healthy, and children need them for growth and health.

Our cardiovascular problems stem from a number of factors—including our relentless consumption of refined sugars, altered fats and our inability to break them down properly. We are so entrenched in the belief that fat *consumption* is the major cause of heart disease that we are blind to the real and underlying reasons.

# HAUTE CUISINE OR OAT BRAN?

The French are an interesting example of another population with an exceedingly high per-capita fat intake—but low cancer and heart disease statistics. In fact, they have the second lowest rate of heart attacks in the industrialized world. World Health Organization (WHO) statistics show that the French consume more butter, cheese, and lard than Americans, and almost as many eggs and meats. According to our current mode of thinking they should be dropping like flies.

The French enjoy their food, take time for meals, and experience a real pleasure in eating. They prefer foods that are whole, fresh, and lightly cooked to those that are packaged, 'heated to death', and processed. Although their diet is loaded with fats, these generally tend to be in their natural state, and the overall mineral content of the food is generally better than that in the U.S. They also have a penchant for organ meats—a rich source of not just fats, but also the trace element, copper. According to Dr. Leslie M. Klevay M.D. at the USDA Human Nutrition Center, "...a lack of this trace-element results in higher levels of cholesterol, blood pressure and uric acid, and impaired glucose tolerance as well as ECG abnormalities."

With the exception of the huge commercial farmers, many of the French country people are still farming the way their grandparents and great-grandparents did, using manures and restoring balanced nutrients to the soil. The soil mineral content is high and the foods and wine-grapes are correspondingly high in minerals. Grape vines also put their roots down very deep into the soil and pull up minerals from deep in the earth.

The French, who enjoy eating a great variety of foods, stay healthier than Americans who are turning their diet into a form of torture, in the name of health. Is the deprivation working?

# WATER AND HEART DISEASE

Minerals dissolved in water exert a protection against heart disease, cancer and other diseases. In 1973, *The Lancet* Medical Journal published a paper showing the correlation between low intake of minerals, especially magnesium, and heart disease. A 1992 study in the *European Heart Journal* looked at 76 communities in Sweden to see if there was a relationship between water hardness and mortality from ischemic heart disease and stroke[15].

The data collected showed that diversity in the water minerals from community to community accounted for 14 percent of the variation in stroke mortality and 41 percent of the variation in ischemic heart disease. At first glance these figures seem improbable, as they only relate to water mineral levels and not to foods which are a much more significant source. The key to this lies in the fact that minerals dissolved in water are far more bioavailable than those present in food. Natural practitioners stress that mineral bioavailability is more important than the amount of minerals consumed, and the results of this study help to confirm it.

Restoration of electrolyte balance, or homeostasis, helps the body and the entire cardiovascular system to stay healthy. It also assists in the normalization of either high or low blood pressure, in conjunction with other treatments, because osmosis, or the fluid pressure and presence of electrolytes on the inside and outside of cell walls, is controlled by mineral availability.

## STRESS AND MINERALS

The body's chemistry is altered dramatically by stress. In the industrialized 'civilized' countries we appear to be more stressed now than at any other point in history. Stress affects body chemistry, creates free radicals and depletes minerals. Hans Selye, a renowned stress researcher, says that, "The-fight-or-flight response increases the level of body chemicals that transmit nerve impulses (electrolytes, trace minerals) thus depleting the body's store." Minerals are needed for the SOD factor, or the antioxidants that scavenge cell-damaging free radicals.

This is borne out by stress and trace-mineral research from the University of Health Sciences in Bethesda, the University of Michigan Medical School, and the USDA Human Nutrition Research Center in Maryland. In the study the researchers monitored plasma levels of zinc, iron, copper, selenium and certain blood proteins in 66 men throughout a 5 day period of physical and psychological stress.

The participants in the study were eating foods that contained more than the RDA of minerals, but despite this, their collective levels of zinc decreased by 33 percent, iron by 44 percent, copper by 12 percent and selenium by 9 percent. In addition, mineral carrying proteins increased as much as 59 percent which demonstrates the removal of trace minerals from tissue stores for use as neurotransmitters. When the stress was reduced to 'normal' levels it took an average of 7 days for the subjects to recover from the mineral deficiencies and register their former mineral levels, despite the 'adequate' dietary intake.

Our reaction to stress as well as our thoughts, beliefs and feelings affect the health of every single cell in the body. In 1980 scientists at Ohio University were studying the metabolism of fats in rabbits who are particularly sensitive to CHD. They provided a high cholesterol diet, and all the rabbits exhibited high cholesterol levels by the end of the study, except one group. To their surprise the researchers found that the only difference between these and the rabbits with high cholesterol levels was that they were taken out, cuddled and spoken to by the technician who was feeding them.

Deepak Chopra, M.D., author of *Quantum Healing*, talks about a study which was set up to look at the risk factors common to those who had heart disease. It was found that the 'standard' risks, i.e. high blood pressure, overweight, smoking, diabetes, etc., did not apply to 50 percent of the people in the study. But they did eventually find two common factors: job dissatisfaction and a low level of personal happiness.

In our civilization more people die on Monday morning than at any other time of the week. The chemicals produced in the body as a result of our thoughts are incredibly powerful, positively or negatively; and the negative ones, or dis-stress, deplete trace minerals.

# OSTEOPOROSIS

*"Artificial agriculture has made us a nation of sicklings in spite of the 'statistics' which claim we are the healthiest nation in the world. These are based on mortality rates which claim greater longevity. This simply means that medical science now enables us to be sick for a greater number of years than formerly."*

❦

Normal Agriculture, November, 1951

## A NEW PERSPECTIVE...

If you read magazines, cereal packages or watch TV then you know about calcium and osteoporosis. You can't help it. Osteoporosis has been gaining almost as much publicity as cholesterol. But the same old advice gets trotted out magazine after magazine, cereal package after cereal package, TV report after TV report—YOU NEED CALCIUM; preferably by downing gallons of milk and, if you are into, or past menopause, by taking calcium and a prescription for hormones—quick, before you disintegrate.

The scare stories have sent women rushing off to the supermarket for low fat milk, and to the medical profession for estrogen replacement therapy (ERT). But we're only being given, and acting on, part of the information.

We're also missing out on what is going to happen when we act on incomplete information. For most people, a lifetime of faithful milk chugging simply isn't going to prevent osteoporosis. I apologize for bursting this bubble. By the time we get scared into thinking we need milk as a calcium supplement the majority of us are past the point of manufacturing the enzymes to digest it properly. And milk's mineral make-up may even unbalance the body's minerals enough to *contribute to* the development of osteoporosis—if you have enough of it for long enough.

## HORMONES

Estrogen replacement therapy is based on the typical medical model—hold off the problem for as long as possible, then when degeneration gets too intense, either operate or assert that the person is "too old" for the problem to matter any more.

During and after menopause the body's ability to hold on to calcium declines, and bone-calcium loss can be intense for as long as ten years after menopause. Without hormonal (or nutritional) intervention the average American woman will have lost up to 50 percent of her bone density, and may have already fractured a hip by age 65–70. By administering ERT at menopause and for up to ten years after, the occurrence of brittle bones can be held at bay for another decade, meaning that women taking ERT might be around 80 years old before a wrist, hip or vertebrae snaps. This is considered a wonderful advancement as we are expected to have expired by then.

ERT is a potent but palliative measure which, for the time being is being touted as the answer to this degenerative disease. Some of the most important studies done on ERT contradict each other and no studies have been conducted for long enough to really prove its safety, long term effects, or its effectiveness. As a matter of fact, the study is in progress right now, the subjects being anyone taking ERT.

In many ways ERT is a beam of medical light, especially for those who are suffering from severe menopausal symptoms. However, its use sweeps the problem under the carpet, so to speak. Having ERT available for the 'crisis' precludes research into natural preventive measures women can take *before* menopause begins—in order to insure against having a reliance on drugs during and after 'the change.'

Let's backtrack at this point to understand estrogen's role in protecting the bones from excessive calcium loss. At menopause the decline in estrogen from the ovaries alters the secretion of hormones from glands like the thyroid and parathyroids, resulting in the cells responsible for breaking down bone becoming more enthusiastic than those that build it.

Before menopause, circulating estrogens reduce the sensitivity of bones to the calcium extracting effects of parathormone (PTH), and stimulate the liver to produce a protein that binds to certain adrenal hormones, decreasing their ability to dissolve bone.

Bone fragility is one of the body's usual, though not entirely normal, capitulations to time, especially in women, and this explains the importance of estrogen and the reasoning behind giving it at menopause. But from my perspective, the philosophy of dealing with the result, not the cause, is a little redundant—after all, shutting the stable door *before* the horse bolts makes a lot more sense—than after.

Osteoporosis is becoming an epidemic for several reasons, most of which correspond to those related to high cholesterol, cancer and other degenerative diseases. If you mix together inadequate mineral and trace-mineral intake, generous quantities of non-foods (white sugar and flour, alcohol and coffee) and those which create imbalance in the body, season this liberally with pollution, chemicals, inactivity and emotional factors including stress, and allow the mixture to simmer slowly over a lifetime, then you have the perfect 'recipe' for disease.

## THE ART OF BALANCE

Gone are the days when families and small communities toiled to produce their own food. Now we rely on the multiple efforts of agribusiness, supermarkets, fast food franchises and the rest of the food purveying alliance to provide our meals. But not one of these giants have the motherly concern for our health that we'd like to think they have, consequently we are eating nutritionally unbalanced and depleted foods—that are unbalancing us.

Osteoporosis, like osteoarthritis, is a repercussion of imbalance. The equilibrium of specific minerals in the blood, including calcium, is critical, and the body has special mechanisms to ensure their constancy. To quote Nancy Appleton, Ph.D., author of *Healthy Bones*, "Calcium is so critical to life, that the body is willing to sacrifice bone mass to ensure adequate calcium in the bloodstream."

Calcium's availability fluctuates all the time, so the bones act as a 'bank' which the body can draw on if dietary calcium is low, or return calcium to, when the absorption is high. Deposits and withdrawals maintain homeostasis and are under the control of hormones that increase calcium release from the bones; its absorption through the intestinal wall and re-absorption through the kidneys; or the deposition of excess calcium in the bones.

Although this system can be disrupted by dieting and inadequate nutrition, excesses of certain non-foods and processed foods, unremitting stress, and long-term use of particular medicines, it is generally not until menopause that the process of bone loss becomes accelerated to a critical point. At this time the balancing mechanisms get disrupted, and for many women the net result is more calcium being taken from the bones than is deposited. This coincides with a decreased ability to absorb calcium.

Bone is a tough, flexible matrix of protein fibers known as collagen, the spaces between which are filled with calcium and phosphorus and small amounts of other minerals such as manganese, fluorine, zinc and sodium. Healthy bone is something like reinforced concrete with the collagen framework providing flexibility and tensile strength, and the minerals providing compression strength.

# GLANDS, HORMONES AND MINERALS

Trace minerals or electrolytes are involved in the health of the glands and the production of the hormones that control calcium's use in the body. Because of this, they should be one of the first considerations for the prevention of exaggerated menopausal symptoms and to reduce the risk of osteoporosis at this time.

The glands are responsible for the production of hormones[16] which regulate the body's functions including calcium and bone density. They can only be healthy when they receive the right nutrients, the most critical being the trace minerals or electrolytes. These minerals are also involved in the actual synthesis of hormones and in the production of sufficient digestive acids to enable calcium absorption in the duodenum.

All the body's functions are sensitive to changes in mineral balance. For instance, if the body is out of homeostasis (disturbed electrolyte balance) calcium deficiency will occur, no matter how high the calcium intake. Everyone popping 'calcium rich' antacids and drinking quarts of milk are doing themselves more harm than good—because they are setting up a mineral imbalance in their bodies.

Milk has a very high phosphorus to magnesium ratio—which restricts absorption of magnesium—which restricts absorption of calcium. It also has a disproportionately high level of calcium compared to other minerals (which is obviously no problem if you are a calf). Too much calcium may disrupt the body's levels of iron, zinc and manganese and these minerals also influence calcium's absorption. If you can't digest lactose you won't get the calcium either.

Certain antacids contain calcium, and the advertisements for them insinuate that they supply *the body* with calcium. Enough people have been duped to have increased the profits from these medicines very nicely[17]. But antacids reduce stomach acidity (that's their function) and calcium needs a high level of stomach acid to be absorbed. Additionally, most of the mineral is going to be used in the process of neutralizing acid when it is taken after meals. Antacids also contain forms of sugar that inhibit calcium absorption. And, probably most importantly, they contain no trace minerals or magnesium, calcium's indispensable partners.

David I. Watts, D.C., Ph.D., in a powerful article on nutrient interrelationships published in the *Journal of Orthomolecular Medicine*, states, "...excess intake of a single element can decrease the intestinal absorption of another element. As an example, a high intake of calcium depresses intestinal zinc absorption." A parallel problem exists with calcium's synergists and antagonists. In order to absorb and use calcium the body needs vitamin D, magnesium and trace minerals including copper. Too much vitamin A or E, potassium, phytic or oxalic acids (from grains and certain vegetables) will upset calcium metabolism and decrease its availability.

Magnesium maintains calcium in its ionized form in the body, keeping it in solution and assisting in its absorption. It discourages the release of hormones that extract calcium from the bones and influences the release and production of calcium depositing hormones.

When minerals are out of solution in the body, and their co-factors are deficient, this may contribute to osteoarthritis and kidney stones. One in ten Americans suffer from kidney stones and the physicians freely admit they don't know why this problem is increasing. Patients are told to avoid calcium, when what they really need is education about diet and mineral balance, trace minerals and magnesium. The latest research shows that the right kind of calcium with its synergistic factors actually helps to *prevent* kidney stones.

## MINERALS AND pH

Now for a surprising twist. Sodium is as important in the prevention of osteoporosis as calcium and magnesium. Not table salt or sodium chloride, but the kind of sodium found in ripe fruit and vegetables. Nancy Appleton, Ph.D., author of *Healthy Bones,* explains that calcium can become deficient because of a sodium deficiency, and a sodium deficiency can occur because the body is struggling to conserve mineral homeostasis (balance).

The blood has to be kept within a very narrow range of pH 7.4 to maintain homeostasis, but the typical lifestyle and dietary habits of people eating the standard American diet (SAD) tends to tip the balance towards an acid pH.

The body has a buffer system that uses sodium to restore physiological equilibrium[18] and its need for sodium in the buffer system intensifies when the diet is loaded with proteins, grains, sugar and refined foods. These all have an acidifying effect in the body which draws on this mineral to maintain a neutral/alkaline pH.

The average diet contains few uncooked vegetables and those eaten are most likely to be commercially grown and low in sodium, compared to levels found in naturally/organically grown vegetables. If there isn't enough sodium around to do the job the body is forced to use calcium as a second buffer. And if the diet doesn't supply much calcium, or if absorption is inefficient, then the bones give up some of their reserve. This demand for calcium as a substitute buffer helps to deplete it from the bones and contributes to the development of osteoporosis.

What you don't eat or drink is as important as what you do. Many so-called foods upset mineral equilibrium and cause extra minerals to be excreted. Sodas (including the diet variety), for example, contain unproportionately high levels of phosphorus which binds with calcium and carries it out of the body. Calcium loss increases following the consumption of either white sugar, salt or caffeine—and combining sugar and caffeine causes calcium loss to peak. Too much salt pulls calcium away from the bones and causes it to be excreted in the urine.

## STRESS

Another surprise element in the development of osteoporosis is the effect of persisting stress. Besides its devastating impact on trace mineral levels, and the creation of free radicals, stress also depletes calcium. I don't mean the kind of stress that gets you to accomplish things, but dis-stress; anxiety about something, doing too much, grief, worry, emotional stress and so on. In these situations (and also when caffeine intake is high) the production of the hormones norepinephrine and epinephrine or adrenaline is increased.

In certain situations a rush of adrenaline can be a life saver, but when stressful responses to 'life' constantly send adrenaline into the bloodstream, the excess plays havoc with the skeletal system—because adrenaline is capable of dissolving bone. The long-term use of the pharmaceutical version of adrenalcortical hormones (cortisone), certain other steroid drugs, dilantin and phenobarbital also contribute to the development of bone disease and disrupt calcium use in the body.

Continuous stress responses result in the over-secretion of insulin which accelerates the loss of calcium from the bones, contributes to the body's acid overload, interferes with mineral balance and minimizes the body's ability to utilize nutrients.

## WARNING SIGNS

Bone loss occurs over several decades and its progression may not be obvious until something extreme like a fracture occurs. Pre-warning signs that calcium and electrolytes are deficient and bone-loss is occurring are often dismissed as minor ailments and ignored, but they can include:

- Brittle or soft nails
- Receding gums, excessive tartar and loose teeth
- Muscle cramps
- Lower back pain (not caused by injury or muscle tension)
- Bone joint and hip pains or aches
- Cramps in legs or feet at night
- Extreme fatigue
- Premature graying of hair
- Heart palpitations
- Glucose tolerance problems
- Uncontrollable temper outbursts

Standard X-rays will detect osteoporosis, but only when there has already been as much as 30 percent bone loss. As a preventive diagnostic tool they really aren't much help. Dentists can warn of impending osteoporosis because the jaw shows bone depletion earlier than the rest of the skeleton and this appears on dental X-rays. Hair-mineral analysis (ask your natural health practitioner) can also give you clues as to how your body is absorbing or utilizing minerals including electrolytes, calcium and magnesium. Even if you think you are eating well, a hair-mineral analysis can let you know your *true* mineral status.

## HIGH RISK

If you are of northern European heritage, fair skinned, blonde (or red) haired, small boned and thin—watch it. Statistically you tend to fare worse than those from a black or Mediterranean heritage with dark skin and hair and large bones. Other high risk factors include:

- Smoking
- History of osteoporosis in the family
- Underweight or anorexia[19]
- Lifestyle with inadequate exercise
- High consumption of refined sugar, alcohol[20], coffee, fats, meat, sodas
- Diabetes
- Rheumatoid arthritis
- Athletic training to the point where menstruation stops
- Early menopause

- Ovaries surgically removed
- Thyroid, parathyroid or adrenal insufficiency or malfunction
- Drinking chlorinated water as main water source
- Long-term use of corticosteroids (cortisone), diuretics, some anticonvulsant drugs, aluminium-containing antacids, and Tetracyclines

## KEY NUTRIENTS

All minerals are important in the prevention of osteoporosis. For instance; studies on rats conducted by Paul Saltman of the University of California San Diego in La Jolla concluded that lack of manganese inhibited the cells responsible for laying down new bone, resulting in bone loss and increased porosity[22].

Soils that are heavily dependent on fertilizers with a high phosphate content have less boron. These fertilizers tend to depress the uptake of this trace mineral from the soil. Boron *decreases* the urinary excretion of calcium and magnesium, increases blood levels of estrogen and testosterone, and is now thought to be necessary for the synthesis and secretion of estrogen, vitamin D and other hormones. Silica (silicon) influences the uptake of calcium in the bones and can be transmuted to calcium in the body. And zinc aids calcium absorption.

Exercise influences bone density. Professional tennis players and pitchers have at least 35–50 percent more bone in their playing arm, and women who lifted weights consistently for 4 years lost only half the bone mass in their arms than would have been expected without that particular form of torture.

Vitamins are extremely important as they can also act as hormone precursors and prevent the loss of minerals in the urine—but it is important to remember that they are *catalyzed* by minerals. Mineral deficiencies result in less vitamin activity in the body. Vitamin K found in chlorophyll helps to cut the loss of calcium in the urine, vitamin E assists glandular function and the body's production and use of estrogen, and vitamin C[23] potentiates estrogen and is essential for collagen production (the main supportive protein of the bone matrix). Certain herbs can also support adrenal function and the production of estrogen after menopause.

As people age they tend to obtain less vitamin D from exposure to sunlight[24] and are inefficient at converting dietary vitamin D into its active form in the body—a hormone known as calcitriol. Some researchers believe it is a lack of this activated form of vitamin D, not so much a lack of calcium that is partly responsible for the current epidemic of osteoporosis. Electrolytes are vital for glandular function and the production of hormones, so again, we need to focus on trace minerals.

Absorption of all nutrients is a major key in the prevention of osteoporosis, including the absorption of protein. Electrolytes adjust the pH of the digestive tract and are the basis for the production of enzymes, including digestive enzymes, which enable the body to breakdown nutrients including protein—and use them. Protein utilization is an important, but little known piece in the osteoporosis puzzle. The body uses amino acids as building blocks, the tough collagen protein fibers form the bone matrix.

There is evidence that some of the bone loss of osteoporosis might be reversed if discovered early enough and the underlying cause is treated. That involves different areas for each person, but it can always include restoring homeostasis, glandular function and amino acid utilization. This is enhanced with liquid electrolytes and a calcium and magnesium supplement in a liquid citrate and aspartate form with electrolytes and other synergistic factors. Research shows that calcium citrate is the most easily absorbed form of calcium (especially when combined with magnesium and electrolytes) for women into and past menopause[21].

Liquid chlorophyll, sunlight, protein, other nutrients and stress-reaction reducing techniques contribute to healthy bones, as do rest and exercise.

And the big one...*stay away from refined sugar.* It inhibits calcium absorption, alters insulin metabolism—which affects calcium metabolism, and messes around with vitamin B6 levels which upsets magnesium...but worst of all, it robs the body of trace minerals so rapidly and thoroughly that it upsets every physical process, and increases the long-term risk of osteoporosis.

# WEIGHT LOSS, BODY BUILDING AND TRAINING

*"I've been on a constant diet for the last two decades.
I've lost a total of 789 pounds. By all accounts I should
be hanging from a charm bracelet."*

❦

Erma Bombeck.

## THE REAL KEY TO SUCCESS...

Crash diets and quick weight loss products work with the symptoms, not the underlying causes of weight gain. And they only have temporary success precisely for this reason. Overweight is not always a symptom of overeating (it is often a symptom of malnutrition) and it's also a symptom of imbalance somewhere in the body. For many people, drastically limiting the diet makes the weight problem worse over time, they progressively eat less and less and still gain weight, despite the feeling that they are starving themselves.

Advertisements featuring svelte people professing to be mere shadows of their former selves (thanks to some awesomely delicious 'wonder shake') are very tempting if you have some weight to lose. But the truth is that people live on these for a few weeks, lose a few pounds, go back to eating meals again—and put the weight right back on, sometimes more than before.

Dieting alone doesn't work for most people. It doesn't work because during the course of drastic caloric reduction the body's metabolism slows to enable it to run on less energy. This means that fewer calories are burned, and even worse, the metabolic rate can remain depressed for up to a year after the diet is stopped. So even a slight, or occasional, increase in calories can lead to an 'out of proportion' gain in weight.

This is what is known as self defeating dieting. Every time the body is put through an ultra low calorie episode it becomes more efficient at surviving on less. When you go back to normal eating (whatever that may be for you) the body suddenly has more calories than it knows what to do with. Well, it does know, it stores them as adipose tissue— fat. And round and round you go on the weight loss, weight gain merry-go-round.

This rebound effect is also attributable to the loss of muscle during dieting and its replacement with adipose tissue or fat when the diet is stopped. Muscle is less fuel efficient than fat and uses more energy. Muscle tissue burns calories even during sleep, and generally keeps the metabolic rate higher. So, the greater the amount of muscle tissue, the higher the metabolic rate—and the more fat stores you'll burn up. Muscle *burns* calories—fat *stores* calories.

Each round of dieting, unaccompanied by the right nutrients and aerobic and muscle-building exercise, results in the loss of muscle, and a subsequent slowing of the metabolic rate after about 10 percent of body weight has been lost.

## *NUTRIENT LOSS*

Diets reduce calories, but they also reduce nutrients—and the body's ability to absorb them. The irony of this lies in the fact that the more nutrients your body gets and absorbs, the faster it will burn off fat. So most diets have an innate self-limiting effect.

Vitamins and minerals often become deficient during a diet, and this is most dangerous in young, adolescent girls as growth and maturity at this time intensifies the need for protein, vitamins and minerals. No matter what your age though, intense dieting and calorie restriction doesn't do your body any good, or your future prospects for maintaining the healthier weight.

The nutritional limitations of diets can have a domino effect with health, especially if the diet is long-term or frequently repeated. The first domino to fall represents the trace minerals which are already scarce even in a good, varied diet. Cutting calories cuts nutrients, and trace minerals get lost in the nutritional shuffle (this is intensified if you are taking certain medications, especially diuretics or cathartics). Then the rest of the dominoes fall…electrolytes adjust and maintain the pH of the digestive tract, when they are missing the healthy gut bacteria can't survive and the unhealthy variety expands.

This results in poor digestion, a predisposition to candida yeast infections and a yearning for sugar and refined carbohydrates, breads, cakes and possibly alcohol—and a further reduction in nutrients available to the body. Repeated consumption of refined foods leads to cravings as the body tries to respond to imbalances in blood sugar and the loss of chromium and electrolytes.

Regular intake of refined sugar can make the problem worse, causing people who are prone to hypoglycemia to overeat in a desperate attempt to extract enough energy to remain semiconscious and feel OK about themselves. Sugar works temporarily as it raises *blood* sugar, but the concentration of empty carbohydrates throws the system out of balance so that it overreacts and brings the blood sugar down too low, generating the roller-coaster energy/lethargy syndrome known as hypoglycemia or low blood sugar. Meanwhile, the body sees all the extra fuel it's receiving as a storage opportunity and most of the empty calories are converted into fat rather than energy.

For some people, what started out as a series of crash diets, consistently starving themselves, or living on a diet of artificially sweetened, no calorie diet foods and sodas—with the occasional 'real' meal thrown in every now and then, becomes a free-fall plunge into blood sugar problems. As time goes on blood sugar control may go haywire and the succeeding 'addiction' to sugar, cake, candy, chocolate or alcohol piles the weight right back on.

Although artificial sweeteners have no calories when tested in a laboratory, as soon as they get into the body's chemistry lab the reality changes. Artificial sweeteners like Aspartame™ are converted to alcohol in the body and can actually enhance the appetite for sugary foods. Excessive refined sugar and carbohydrates like white flour products deplete chromium, which is then unavailable to enable the body to utilize the refined foods properly. This begins to set up blood sugar problems and the storage of glucose as *fat* rather than its conversion to energy or its contribution to muscle development.

## LESS MINERALS—MORE WEIGHT

The trace mineral chromium is scarce in the standard American diet due to deficient soil and food refining. Dieting can cut its availability to almost zero.

Chromium aids the body's blood sugar metabolism and fat burning. Acting as a co-factor to insulin, chromium is involved in energy production, muscle development, fat and cholesterol metabolism and the regulation of blood sugar. It also allows the body to hold on to more muscle during calorie reduced diets by promoting the entrance of glucose and amino acids into the cells to make muscle. When glucose is being used to build muscle far less of it is left floating about for the body to turn into fat.

Biologically active chromium helps in the reduction of sugar cravings through its action with insulin and blood sugar, it also helps insulin transport the amino acid tryptophan into the brain for serotonin synthesis. Serotonin is one of the neurotransmitters responsible for suppressing the body's desire for carbohydrates.

The popular diets don't provide the body with enough trace minerals or bioavailable chromium and for this reason Dr. Jeffrey Bland encourages trace element supplementation whenever food is limited or 'meal replacement' powders are used. He notes that minerals work together and the body has to take in (and be able to use) minerals such as manganese, zinc, copper, and selenium in order for chromium to be active.

## IRON WILL, OR IRON?

Most women on long-term calorie restricted diets or on a self-induced starvation regimen do not get enough iron. At best the 15-18 mg. of iron a woman in her teens through her late 40's needs is hard to obtain from 3 meals a day— on a restrictive diet it is almost impossible.

Iron deficiency is the greatest single mineral deficiency worldwide, and most women are walking a thin line between sufficiency and deficiency. A diet can tip them over into deficiency because iron availability is reduced.

Most iron-rich foods are high in calories and in order to meet the daily requirements as many as 2000 calories may have to be consumed. Only about 5-15 percent of the iron in food will be absorbed, and when dieting, iron's co-factors, nutrients like vitamin C and trace minerals (which aid its absorption) also tend to be scarce. A rigorous exercise schedule or regular aerobics added to the dieting will further deplete iron.

A deficiency of iron produces fatigue, lethargy and lack of stamina accompanied by a general seizing up of the mental equipment and a feeling that the legs have turned to lead[25]. The natural assumption is that the reduction of food on a diet is causing the problem and the dieter either subsists at this level until the self-imposed regimen ends, or starts eating and eating in the hope that the mental faculties will return and the energy will improve. Depending on the level of caloric restriction and the type of food being eaten, the symptoms may well be a deficiency of iron, not necessarily a deficiency of calories. The body is crying out for more trace minerals including bioavailable iron, more vitamins and unrefined, non-phytic acid foods. The usual response is the consumption of more empty food— and further weight gain.

Many changes take place in the body from the diminished nutritional state after a prolonged and drastic diet. There may be increased sodium and water retention (edema), increased insulin production (lowering of the blood sugar), and altered intracellular enzyme activity and breakdown rates of muscle and adipose tissue. Cardiac arhythmia, (abnormal heartbeat) experienced by some people during a drastic bout of dieting may also have its roots in deficiencies of essential minerals.

## *PROTEIN*

Most nutritionists suspect that the average person assimilates no more than about 20 percent of their protein intake—and diets make this situation worse by depleting nutrients including electrolytes that are needed to metabolize proteins.

Without trace minerals protein metabolism suffers and this affects many physical processes. The trace minerals in electrolyte form have a diuretic effect on excess water in the body by regulating protein metabolism and maintaining internal water balance. A lack of protein or an inability to utilize it well can result in retention of water. More fluid is drawn into the cells which then hold onto it. Cravings for sugar and sweet foods are also related to protein deficiency. But the antithesis, filling up with protein or large amounts of separate amino acids unaccompanied by electrolytes, saturates the body with harmful waste products. This is caused by the incomplete conversion of protein to amino acids. In this situation the body creates uric acid instead of new tissue, and is then forced to use more minerals trying to neutralize it.

Proteins are used to make cells and tissues and for building and maintaining the body. Amino acids, the components of protein are the 'building blocks' of life and have special functions as components of hemoglobin, plasma, antibodies, hormones and enzymes.

Proteins and electrolytes also help in the exchange of nutrients between cells, the intracellular fluids, and between tissues and the blood and lymph. The enzymes needed for digestion are created from amino acids with the help of electrolytes. Minerals enable proteins to be broken down into their component parts (amino acids) which then become bioavailable—available for the body to use.

Cutting calories too drastically forces the body to use its internal reserves of fat and protein to meet energy requirements. On a short-term diet program it adapts to the caloric deprivation by using muscle protein and cellular fats as the primary source of calories. After a couple of weeks the body begins to undergo extensive protein loss.

Protein replenishment is vital as body proteins are always changing and it cannot make them itself. Because of their continuous turnover they must be supplied in every meal, and need to be broken down efficiently into amino acids. Trace minerals make this possible.

Trace minerals and amino acids (components of protein) work together. What really counts in this instance is the quality of the amino acids available to the body, not necessarily the quantity—and the body's ability to assimilate and utilize them.

Trace minerals are needed for the proper breakdown of proteins through the production of enzymes. A full spectrum of high quality amino acids coupled with electrolytes or trace minerals enables the vital elements to become available to the body instead of poisoning it.

## STRESS AND WEIGHT

Any diet involves a certain amount of stress, and a lack of electrolytes makes it harder for the body to neutralize its negative effects. Stress is a big factor in weight problems, 79 percent of women questioned in a poll said that stress was the main reason for their overeating. Other research suggests that 'yo yo' dieting puts so much stress on the body that it may put people at risk of early death. To understand how stress affects the body let's compare an ancestral stress response with today's.

Imagine a caveman quietly picking berries in the woods. Suddenly a very large, hairy, hungry carnivore appears on the path. Faced with extinction the neanderthal experiences stress which sets off a chain of events in his body that we call the fight or flight response. The brain sends urgent signals to the adrenal glands which respond by releasing a hormone called adrenaline. This rushes the body's fuel (blood sugar) out of the liver and into the blood stream, raises breathing, pulse, blood pressure and heart-rate and tenses the muscles. In a matter of seconds the caveman can run or fight with Olympian speed or strength.

Today, as then, whenever stressful circumstances occur, the same chain of events takes place in our bodies. For some people the stress response is a manageable event. In certain situations a powerful stress response can be lifesaving (a mother seeing her child trapped under a car is suddenly able to lift the vehicle), but sometimes our antique stress response, especially if repeated several times during the day, is not healthy.

When your boss calls you in 'for a chat,' your co-workers or customers are particularly cantankerous, you get stuck in a traffic jam, or catch sight of blue flashing lights in your rearview mirror, society dictates that it isn't appropriate to attack, or run away screaming. Any of the above actions would have helped to dissipate stress chemicals in the body, but because you are 'civilized' you boil away quietly inside and experience different levels of symptoms like palpitations, dry mouth, knots in the stomach, tension and rapid breathing.

The body's chemistry is dramatically altered by stress which creates free radicals and depletes minerals. As I mentioned earlier, Hans Selye, a renowned stress researcher, says that, "The-fight-or-flight response increases the level of body chemicals that transmit nerve impulses (electrolytes, trace minerals) thus depleting the body's store." Minerals are needed for the SOD factor, or the antioxidants that scavenge cell-damaging free radicals.

Even mild stress reduces vitamin C and this adds to protein loss during a diet. Humans, guinea pigs and apes do not produce their own vitamin C in the body but most other animals do. Researchers found that when they stressed goats they produced literally thousands of milligrams of vitamin C in response. Vitamin C or ascorbic acid, together with electrolytes, assists in the neutralization of damaging free radicals.

Stress also results in an inability to digest food properly directly depleting trace minerals, macro-minerals and vitamins, all of which are needed to neutralize its negative effects!

Stress lowers serotonin levels in the brain and scientists think that decreased levels of this neurotransmitter are associated with feelings of hunger. High levels may be associated with feelings of satiety after eating. It is believed that people who are overweight tend to have lower than normal levels of serotonin.

## STRESS AND BLOOD SUGAR

Constant stress intensifies the problem of low blood sugar during a diet because it overworks the adrenal glands. They are bullied into secreting excess adrenaline which, in turn, causes sugar stored in the liver (as glycogen) to be dumped into the bloodstream. The pancreas then has to overreact and bring the blood sugar down.

This scenario is intensified by the use of tobacco and the consumption of coffee and simple carbohydrates like sugar.

Consumption of complex carbohydrates, as opposed to simple carbohydrates, is believed to encourage the production of serotonin by increasing the uptake of tryptophan (serotonin's precursor) in the brain. It also alters the physical experience of hunger.

The sugar or glucose molecules in complex carbohydrates are linked in an intricate, branched pattern. It takes the body more time and effort to detach these molecules from the chain than it does from simple carbohydrates like white sugar or flour, so glucose is released into the body slowly leaving a feeling of satiety for longer and decreasing the craving for carbohydrates at the next meal.

Unlike refined sugars, complex carbohydrates from whole foods, will not send energy crashing an hour after their ingestion. The body burns more calories processing proteins and complex carbohydrates than it does burning fat.

Our response to stressors as well as our thoughts, beliefs and feelings affect the health of every single cell in the body, and the coinciding loss of trace minerals can alter body systems so much that it becomes almost impossible to lose weight and keep it off.

## TOXIC WASTE

When people are told to lose weight, the health of their digestive system, eliminative system, their pancreas, liver, adrenals and thyroid all get lost in the idea of the diet. The emphasis is on cutting down the food taken in, not on supporting the health of the systems that work to keep the body at its ideal weight and function. A drastic diet not only starves the glands and organs, but it can also overwhelm them (especially the liver) with toxins.

Several pounds of overweight represent several ounces of concentrated, stored toxins in addition to the fat. After the first week or so of a drastic diet, tissues are forced to give up a lot of things they've been hanging on to. When the body begins to burn up its stores, chemicals, dioxins, pesticides, metabolic wastes, toxic heavy metals and anything else that was stored by the body gets dumped into the bloodstream along with the stored fat.

Christopher Bird and Peter Thompkins write that, "The activities of enzymes are extremely susceptible to chemicals. Enzyme reactions are influenced by a deficit of any functional nutrient." Dieting cuts down trace mineral availability, upsets digestion and enzyme production, and starves the liver of the nutrients it needs to detoxify the outpouring of chemicals from the tissues.

The chemicals released from fat stores also represent a risk factor for potential metabolic toxicity and can reduce the efficiency of the conversion of food to metabolic energy—leaving the dieter tired and with a toxin induced headache.

Put into this perspective, it begins to make sense why there is often a rebound effect after some diets. The body is starving and the glands and organs have been overtaxed and stressed. As body systems are not functioning efficiently it is not at all surprising that the weight piles back on. What is the outcome of continuous dieting, and what impact does it have on future health or the body's ability to achieve and maintain a healthy weight? I believe that any diet which pays no regard to the health of the glandular system, the liver, or to nutrient absorption should be labelled with a surgeon general's health warning.

## FOOD ALLERGIES AND ADIPOSE

There is a natural tendency to eat less *variety* when dieting. This can set up the possibility of developing food allergies, especially if the diet is ongoing or frequently repeated. Anything we eat every day can be a potential food allergy waiting to happen. Foods that we have become allergic to are not broken down properly and the nutrients are not absorbed well, so the body stores them as fat.

Cow's milk is a prime example. Many of the meal replacement diets involve dissolving powder in milk, which then represents two of the main meals of the day for as long as the diet is continued. Most adults have trouble digesting milk, especially when it becomes the main component of the diet, and particularly when it has been pasteurized, homogenized, skimmed and otherwise mistreated. Trace minerals and acidophilus will help milk digestion, but they both got their 'pink slips' shortly after the diet started.

The solution to this problem is to become aware of, and maintain variety in the diet. The key to staying healthy and keeping excess weight off is a variety of balanced whole foods—the two factors crash diets never promote.

## EXERCISE

Exercise helps to break the "rebound effect." It is also the crucial factor in helping your body to burn off stored energy, or fat. The awful truth is that just one pound of body fat is a source of approximately 3,500 calories of stored energy. If you have 10 pounds or more to lose that represents a lot of not eating. Bicycling for 15-20 minutes burns off about 100 calories so does walking or raking leaves for 20 minutes.

Any form of exercise which speeds up the heart and pulse rate and causes deep breathing revs up the body's energy burning capacity. When done on a consistent basis, exercise burns fat and it also sensitizes cells to various hormones, such as insulin and thyroid hormones that help burn calories more efficiently. To keep this positive, remember to replace electrolytes lost in perspiration.

## SUCCESS IS NEVER HAVING TO SAY YOU'RE DIETING

Successful normalization of weight comes from a change in lifestyle and often a change in beliefs. Thinking about what it is you *do* want instead of what you *don't* want is important. Having a picture of how you will look when you have achieved your goal, talking to yourself about what you want, and feeling how you will feel when you are as you want to be—definitely helps you achieve your target.

Ask yourself if there is a *positive* intention for the weight, and what will happen when you lose weight. Are you comfortable with your answers, do you feel safe? If your answers are tinted with fear, or you recognize a block to your goal, working with an NLP practitioner will enable you to work through issues which may have prevented you from moving forward.

Achieving your desired weight occurs most easily as part of an ongoing holistic approach to life and health. Nothing is going to achieve healthy weight loss on its own. Simply cutting back on calories every now or then isn't likely to do it, exercise alone isn't going to do it, protein shakes instead of proper meals or supplements taken with an abominable diet...nothing short of near starvation or an exercise regimen designed for a 'boot camp' is going to do it alone.

Normalization of weight has to be a combination of different factors—a new life style, a new outlook, new beliefs about yourself, your body, and your abilities. In addition, it helps to understand how important wholesome, live foods and nutrients (especially electrolytes) are in assisting your body in the process of acquiring and maintaining its ideal weight.

Are you packing more calories or energy into your body than are being used or burned off? How is that happening—are the foods you are eating mostly refined empty calories? Is a blood sugar imbalance involved; do you skip breakfast, get a slump at 11:00 A.M., pig out, feel better and then remain semi-comatose for the rest of the afternoon? Is your digestion and glandular system happy—are nutrients being absorbed or is the body constantly hungry for them? Are you missing trace minerals?

There really is too much emphasis on dieting and not enough on diet, which incidentally comes from the Greek word diaita meaning a 'way of living.' When you are basically healthy and make a wholesome diet part of your healthy life style you'll probably never have to bother with dieting again.

## BODY BUILDING AND TRAINING

A colleague of mine is convinced that using steroids for the muscle factor in body building and for enhanced athletic performance—makes men sterile.

Digging into the research on trace minerals has provided me with several clues regarding the importance of minerals in the building of muscle, and how the loss of minerals can reduce sperm count and cause a number of 'sexual malfunctions.'

We know that the prescription steroid, cortisone, upsets mineral homeostasis and that long-term use can lead to softening of the bones. Although the data is sparse on the interaction of drugs with trace minerals, whenever macro minerals are depleted by a drug we can assume that trace minerals are being depleted too.

Taking sex hormone (anabolic) steroids into the body on a long-term basis probably affects mineral homeostasis too, and even if it doesn't—sweating while pumping iron, working-out or training certainly will.

Zinc is just one of the trace minerals lost during a workout. This mineral is present in seminal fluid and is used by the body for the health of the entire reproductive system, and for the production of a normal level of healthy athletic sperm. In men with a low sperm count the replacement of lost electrolytes including zinc, brings the sperm count and their motility back to normal in a relatively short time—if there is nothing else causing the disorder.

Impotence can also be associated with mineral deficiencies. It is thought that a significant number of men with this problem have atherosclerosis of the penile arteries, caused by a diet too high in certain fats and sugars, but most importantly, too low in trace minerals including chromium and manganese. Narrowed blood vessels restrict blood flow and limit erection.

Bioavailable chromium and manganese enable body systems to digest and excrete fats before they can become deposited on artery walls. Trace minerals in liquid electrolyte form also keep minerals in balance so they won't become part of the arterial plaque.

# QUALITY NOT QUANTITY

Proteins are not well assimilated if electrolytes or trace minerals are deficient, and the high protein content of many body building or athletic regimens draws on trace minerals which are already likely to be depleted by the workouts, sweating and high body stress.

Loading up with protein powders and pumping up the proteins from meat creates uric acids in the body which then draws on electrolytes and macro minerals to neutralize them.

Trace minerals and amino acids (components of protein) work together. The quality of amino acids available to the body, not necessarily the quantity, and the body's ability to assimilate and utilize them, is the key to building new tissue and muscle.

Trace minerals are needed by the body for the proper breakdown of proteins through the production of enzymes. Supplementing a body-building diet with high quality amino acids matched with bioavailable electrolytes or trace elements enables the vital protein elements to become available to the body for the muscle factor.

Nature intended electrolytes as the ultimate builder of glandular strength and hormone production. An athlete who understands and uses trace minerals with other key nutrients doesn't need to damage the body with steroids in order to improve performance or build muscle. The desired effect can be achieved in a steady, safe way with the inclusion of a bioavailable formula of trace minerals.

When minerals in electrolyte form are used in combination with an athletic or body building nutritional program they make it possible to achieve the desired muscle factor—because no vitamin, amino acid, carbohydrate or even the body's own steroids are used efficiently without trace minerals.

# BALANCING MINERALS

Mineral balance is critical especially for the athlete. For example, if copper is in short supply, iron utilization will decline and produce symptoms like fatigue and lack of stamina. In addition, the high volume oxygen intake during athletic exertion oxidizes blood cells faster than normal and increases the chance of anemia.

A high intake of meat and other proteins leads to an increase in the metabolic rate, part of which is achieved due to calcium and magnesium excretion. Loss of both these minerals can result in cramping, spasms and irregular heart beat. Iron deficiency interferes with the formation of special enzymes in the body that affect muscle function.

Young female athletes are especially prone to mild anemia or iron deficiency. This can limit speed, stamina and strength and may be responsible for preventing the improvement of all three, even with a regular and disciplined practice program. Excessive physical activity often leads to the loss of menstruation which, in part, stems from the body's depletion of minerals and its attempt to hang on to those which remain. Routine tests for iron deficiency may not show problems, especially if hemoglobin levels are normal. Iron depletion is common and often goes undetected unless serum ferritin levels are measured.

The building of muscle and the production of energy draws on scarce chromium reserves. Acting as a co-factor to insulin, chromium is involved in energy production, muscle development, fat and cholesterol metabolism and the regulation of blood sugar. It also promotes the entrance of glucose and amino acids into the cells to make muscle.

Most diets lack chromium because the soils are deficient in this mineral. Athletes may become deficient in many minerals, especially if the diet consists of refined foods and electrolytes are not replaced after workouts.

In the summer months I see many patients who complain of having felt dizzy or faint half-way through a run, bike ride, tennis game or other physical activity. When questioned, their symptoms usually relate to lost potassium and trace minerals. Taking electrolytes before and after exercise prevents the problem from recurring.

Many serious cyclists will remember the tragedy of Tommy Simpson's death in the Tour de France in 1963. He collapsed and died on a mountain ascent. He was a professional who was experimenting with drugs to increase his athletic performance. He was also under a great deal of stress in addition to fierce heat which had caused him to lose electrolytes in sweat.

Back in the 1960's cyclists cared for their bicycles better than their bodies. Athletes were rarely given sound nutritional advice and very likely to be given salt tablets. Too much of any mineral will unbalance the others. I wonder if an electrolyte formula in Simpson's water bottle and good general nutrition could have prevented his collapse from being fatal.

## THE MINERAL ADVANTAGE

Today athletes and body builders are extremely knowledgeable about their bodies and about many principles of nutrition. They study whatever they can find in order to give themselves the edge over their competitors.

Not much has been written about the importance of trace minerals for body building or athletic performance. However, the information throughout this book points to the importance of minerals in obtaining and maintaining health—for an athlete or a serious body builder electrolytes are an unrecognized key to superior health, fitness and physique.

Electrolytes are also the key to future health—people who devote their early life to a serious or professional athletic career or hobby often suffer later in life from the stresses and sheer physical demands put on their bodies. Significant minerals and trace minerals are lost during workouts and training. The body needs them for maintenance and repair, and to make the SOD factor which counteracts the free radicals produced by intense physical activity.

For any athlete or body builder, trace minerals or electrolytes need to be the first nutritional consideration. They work in the body to produce good assimilation and metabolism of nutrients and the breakdown of proteins into amino acids. Minerals influence the production of energy, play a major role in hormone production and utilization, get involved in the building of new tissue, keep bones, joints, tendons and ligaments strong and flexible, help to keep the entire cardiovascular system healthy, quench damaging free radicals, and maintain the strength of the cells and their fluid levels.

Everything taken into the body needs trace elements to make it work. The first consideration for any serious body builder or athlete should be trace minerals in the form of liquid electrolytes. Anything else comes second when you want to be number one.

# WINNING YOUR MARBLES BACK

*"Do you think your mind is maturing late,*
*Or simply rotted early?"*

Ogden Nash

## *FROM SOIL TO PSYCHE...*

Was it Einstein who said that we only use 10-36 percent of our brain potential? I don't remember, I must be getting senile. Actually, I know I'm fine, but it's common to hear people convincing themselves they're on the road to senility or Alzheimer's disease because their short-term memory has temporarily failed them.

Forgetfulness is quite a common complaint. I don't have the figures from the lost property office of the London Underground (subway system) and the London Bus system, but I know that they collect hundreds of thousands of umbrellas, wallets, pocketbooks, eyeglasses and other personal clutter every year. Most of which remains unclaimed, presumably because the people who lost them forgot where they lost them!

It's not just the British either; a Japanese newspaper reported that passengers on the National Rail system leave an average of $30,000 on trains every day. In a 3 month period 441,000 umbrellas, 396,000 articles of clothing, 72 sets of false teeth and 7 boxes of human ashes were left on Japanese trains. How do you forget your teeth?

## *TWO GRAPES SHORT OF A FRUIT SALAD*

In 1976, when I was 19, I came to the US to visit my Aunt. I had a list of physical and mental complaints that my physician in England had dismissed as part of the hazards of being a teenager. Amongst an array of problems, I was unable to remember telephone numbers, people's names or even what I had gone upstairs for. I spent my life going round in circles getting nothing done.

My aunt took me to see her naturopathic physician. He asked me many questions and gave me a list of minerals, vitamins and herbs to take, besides suggesting that I give up all my favorite foods including cheesecake. I was mortified and thought very seriously about tickets for the next plane home. But within three days of what I considered to be a torturous regimen (I wanted my cheesecake),

I took my glasses off and have never worn them since. Within about a week I realized that I actually liked myself, and within ten days or so I could remember numbers, names and even what I was supposed to be doing...that was an incredible feeling. I had my marbles back.

My mother always uses natural remedies and fed us well, kept us away from white flour, white sugar and, as she describes them, 'poisonous fluorescent ice creams.' Then I began work and started eating the 'food' in the cafeteria at the office—especially the desserts. Over time my desire for sugary foods intensified until I was eating 3 desserts every day (when my mom reads this she'll be horrified). I didn't think about what I was doing to my body, and especially never considered the effects on my brain and psyche, although I knew better. I was not alone; the majority of people don't link their love of sugar, cheese cake, chocolate or white flour goodies to loss of memory, depression, or lack of *joie de vivre.*

When you eat refined sugar it rapidly enters the bloodstream, and the pancreas reacts by sending insulin, a hormone that brings blood sugar levels down, into the blood. By the end of a day filled with sugary snacks and desserts the blood sugar has peaked and dived many times and the pancreas and adrenal glands probably feel as if they've gone through the equivalent of a three hour Jane Fonda workout. When the blood sugar (blood glucose) takes a dive, that's when the psyche and the memory suffer—because the brain's fuel is primarily glucose.

Running parallel with the blood sugar imbalance is an unbalancing of the minerals. Refined carbohydrate foods like sugar and white flour are almost totally devoid of trace minerals and fiber. This means that sugar is absorbed into the bloodstream too fast. Refined sugar is also highly acidifying, and in order to neutralize the acids trace minerals, especially chromium, are drawn from the body's reserves like a pump. This loss of minerals decreases the body's ability to deal with sugars!

Nathaniel Mead, writing in the September/October issue of *East West Journal* in 1991 reports, "Severe depression which strikes one in 50 American teenagers has been linked with single deficiencies of the trace minerals copper, molybdenum, vanadium and zinc." It has also been linked to the macro minerals calcium, iron and magnesium. He continues, "In part, these effects stem from the elements' interactions with other nutrients. People lacking copper and chromium are less able to regulate their blood sugar levels. Copper helps the liver store glucose, and chromium boosts the action of insulin."

In U.S. Senate document number 264 it states, "Many

backward children are 'stupid' merely because they are deficient in magnesium. We punish them for our failure to feed them properly. Certainly our physical well-being is more directly dependent upon the minerals we take into our systems than upon calories, vitamins or upon the precise proportions of starch, protein or carbohydrates we consume."

## MINERAL TRANSMITTERS

Minerals, especially electrolytes, are also an integral part of the 'soup' that carries chemical messages in the brain and nervous system. The human brain contains billions of neurons (nerve cells), each of which can connect to others, but are separated by minute gaps called synapses.

Electrical neurological impulses travel by leaping across the cell-to-cell gaps, and that's how the connections are made and our brain and nervous system functions. In order to perform these biological acrobatics, the signals need a helping hand from substances called neurotransmitters and these are made from, and stimulated by, electrolytes and macro-minerals, amino acids, choline and other nutritional factors.

A naturopathic colleague who lives in Florida was hit by lightning a few years ago. He was standing in his yard, leaning against a metal pole when lightning struck near him. It travelled along the ground, connected with the pole and riveted him to it for several seconds. His dog who was standing beside him was knocked down. Looking back, he thinks he was unconscious for some time and then remembers going indoors. The house had been hit and the electrical system was fried. The bolt was so powerful that it had even caused the guts of the motion detector in the alarm system to burst out of the casing. He was so concerned with the electrical problems in the house that he completely forgot he had been hit.

Within 24 hours he started to have extreme symptoms similar to chronic fatigue syndrome, but didn't connect it with his lightning strike—which he'd forgotten. Getting out of bed became a Herculean task, he couldn't function properly, was depressed, unable to concentrate and had severe headaches. Much of his time was spent sleeping in chairs in a semi-upright position in an attempt to relieve the pain in his skull. He eventually sought advice at the Mayo Clinic and the physician pinpointed the lightning strike but could only offer painkillers. He was told he had irreversible damage to the brain and nerve cells.

He treated himself with electrolytes, macro-minerals, lecithin and phosphatidyl choline, fatty acids, B complex vitamins, wheat germ oil, and numerous antioxidants. It took time, but he is now free of the major symptoms.

Physicians may not know about, or believe in, the ability of electrolytes to aid in neurological repair, but scientists do know that electrolyte, or trace-element, deficiencies can influence episodes of depression, anxiety, decreased attention span (attention deficit disorder [ADD] in children), and poor short-term memory.

## TRACE AND TOXIC MINERALS AND MEMORY

Serious age-related disorders like deterioration of brain tissue, or senility, are linked in several ways to electrolyte imbalance. Besides their role in blood sugar metabolism and neurotransmitter production, trace minerals are involved in the function of the minor blood vessels and capillaries in the brain, as well as the absorption of amino acids (proteins) which the brain needs.

The more available and usable protein there is for the brain, and the more efficient the circulation and transportation of oxygen, (the brain uses 25 percent of the body's total oxygen intake) the better the memory stays over the years.

Scientists have been arguing for years about aluminum's link to senile dementia or Alzheimer's disease[26]. It is known that the brains of Alzheimer's patients contain abnormally high levels of this mineral. Aluminum has an enzyme inhibitory potential and, recent research shows that aluminum deposits prevent the brain from using vitamin B12, which, in turn, prevents it from making acetylcholine—even when there is choline in the diet. A lack of acetylcholine allows the brain to accumulate a nerve hardening and strangling protein called beta-amyloid (*Science*, October 9, 1992). Alzheimer's disease is characterized by a degeneration and tangling of the brain's nerve terminals.

Some Alzheimer's patients become aggressive, and a study by Elizabeth Reese of the Mineral Laboratory in California may give a further clue to the aluminum/Alzheimer's question. Her research involved hair analysis tests on aggressive boys in a detention home. Every one tested had abnormally high aluminum levels.

Interestingly, the similarities between the homeopathic drug 'picture' of aluminum (the symptoms produced in healthy people taking potentised dosages of aluminum) and the symptoms of Alzheimer's patients are the same.

I also have my own theories; based on work by Dr. G.H. Earp-Thomas, we know that plants grown in mineral deficient or imbalanced soil can take up high amounts of toxic mineral elements. The pH of acid rain and artificial fertilizers results in the liberation of aluminum that is normally locked in soils and rocks. It seems probable that some food crops and water supplies contain high levels of aluminum, while the protective trace minerals which stop us from absorbing toxic minerals are dwindling.

Not only are soils lacking trace minerals including zinc, but also macro-minerals like calcium and magnesium. Sufficient intake of all these minerals stops aluminum from accumulating in the brain. Our bodies react a little like the plants and need a balance of minerals—otherwise excessive levels of toxic minerals can be taken up by cells and tissues.

Aluminum only came into general use around 1886. Since then we have expanded its applications and include it in food and water processing (there are worries that fluoride and chlorine in drinking water act as catalysts, speeding up the body's absorption of aluminum), medicines especially antacids, baking powders, deodorant and antidandruff formulas, pickling solutions and anti-caking agents. So, from sources in food, medicines, and cosmetics, and leaching of aluminum into drinking water and soils, we are getting an overdose of a mineral considered to be a nonessential element. On top of all this we are being deprived of the minerals we need to protect us against the overdose.

When the 'experts' say that aluminum is safe and can be ingested in so many parts per million every day, they are discounting the numerous hidden sources, the effects of combination chemical and mineral elements and the decline in protective minerals. They are also talking about a mineral that accumulates in the lungs, liver, thyroid and brain. And a mineral that, in excessive amounts, is a *known* neurotoxin, or nerve poison.

We ingest approximately 5-10 mg. of aluminum each day from foods, food additives and medicines according to British Ministry of Agriculture figures. Although the amount we take in from water is nowhere near a fraction of that from foods, aluminum dissolved in water, like any mineral, has a far higher bioavailability.

What a refreshing change it would be if the people who are supposed to protect our health ceased their defensive claptrap and spent the time gained in researching the protective trace elements and their ability to prevent the body from taking up toxic trace elements like aluminum.

# ALUMINUM UPTAKE

Since time immemorial we have taken our water from running streams and rivers, but in the last century most people have come to rely on city water. This imitation of the 'real thing' has had most of the minerals processed out of it and has usually been recycled and treated with several chemicals, including chlorine, fluoride and aluminum.

Just as crop mineral levels can be unbalanced when soils are low in trace elements and high in aluminum, so too can the water we drink. When acid rain falls into reservoirs, lakes, and on the surrounding land the pH, or acid/alkaline balance, eventually changes, becomes more acidic, and aluminum from the soil and surrounding rocks is leached into the water. This impacts on the health of humans and animals drinking the water, because a high aluminum content destroys the colloidal aspect of water making trace and macro-minerals even more difficult to absorb—but not the aluminum.

A study conducted by epidemiologist C.N. Martin, Ph.D., which was published in *The Lancet* medical journal, January 14, 1989, found the risk of Alzheimer's disease to be unusually high when drinking water contained more than 11 mcg. of aluminum per liter of water.

In the 1800's when Dr. Samuel Hahnemann experimented with small doses of highly toxic substances, he ground them into powders and diluted them in water. He found that this gave them a much higher bioavailability, even though the actual concentrations of the materials were minute. Millions of patients using homeopathic remedies have demonstrated that the body absorbs 'diluted' materials very easily. This is as true for essential trace minerals dissolved in water as it is for the toxic ones.

The body also selects minerals from food and water sources, based on what is and isn't present. Martin Fox explains, "If you have sufficient calcium, magnesium and other essential elements present in drinking water along with lead and cadmium, there is a higher probability that the essential elements will be absorbed and not the nonessential. In this way the essential elements can provide a protection against the absorption of nonessential and potentially harmful elements. Further, if lead and zinc are both present in the drinking water, zinc would be selected by the cell and protection against lead would be provided. But in the event that there is no zinc, lead would be selected..."

# DANGEROUS MISTAKE

The majority of British water companies fail to meet the European Economic Community EEC standards for aluminum contamination. Some drinking water contains hundreds of times the EEC limit for this metal and many add aluminum sulfate during water treatment.

A few years ago, in a small town in southwest England, the drinking water was accidentally contaminated with aluminum. This particular town clarified its water with aluminum sulfate and the trucker delivering the material dumped the entire load directly into the water supply. The accident wasn't made public until several days later.

Physicians began seeing residents who were complaining about vague pains, joint and muscle aches, weakness, sleeplessness, diarrhea, vomiting, skin rashes and mood changes, and many children were unwell. Physicians who treated the children reported that once docile kids had become disruptive and were tearing their surgeries apart.

There were many law suits filed, but the water authorities and the government denied that aluminum could cause any health problems. To date nothing has been resolved and no one is sure how the aluminum will affect the future physical and mental health of Camelford's population.

# ADD, HYPERACTIVITY & INTELLIGENCE

Once parents have succeeded in peeling their hyperactive kids off the ceilings, they ship them off to the medical profession and, more often than not, come back with a prescription for Ritalin. A school principal told me that she estimates as many as 50 percent of the kids in her school are taking this drug.

There is now a shortage of Ritalin. Doctors need to be concerned about the lack of minerals, not the lack of drugs. But as usual we have it backwards.

Nathaniel Mead's article 'Food For Thought' in the Sept/Oct 1992 issue of *East West Journal* pointed out the importance of electrolytes in treating children with depression and hyperactive behavior. He noted that the addition of electrolytes to the diet brought the children's behavior back to normal.

A study to test the significance of vitamins and minerals in intelligence was reported in the Lancet in January 1988. The researchers asked 90 school children aged 12-13 to keep a dietary diary for 3 days. They found that in most cases the intake of vitamins was close to the RDA, although for a minority the intake was low. For most of the children the RDA of minerals was not being met. For the next 8 months either a placebo or a multivitamin and mineral supplement was administered double-blind to 60 of the children. The supplemented group, but not the placebo group or the remaining 30 who took no tablets, showed a significant increase in nonverbal intelligence.

We know that minerals, especially in the electrolyte form, act in the transmission of brain and nerve impulses and are vital for maintaining a placid disposition and sharp intellect, we also know that certain trace and macro minerals prevent the uptake of toxic trace elements like lead which decreases mental ability. Missing trace elements, including zinc with its synergistic factors, and B complex, excessive toxic elements like aluminum and lead, as well as low blood sugar may well be the key in the 'mystery' of ADD and hyperactivity.

Minerals are the delivery system for the brain's fuel and when unbalanced by deficiencies can actually cause problems. Just one example:` it is known that zinc helps to 'control' the levels of copper and manganese and that excessively high levels of manganese have been shown to be a factor in violence. Balancing the minerals will turn out well balanced human-beings.

## BRAIN DRAIN

For many different reasons we are not getting the trace minerals we need. Every cell, tissue, gland and organ in the body demands electrolytes for optimum function and to protect itself against free-radical damage. The brain and nervous system are no exception. The seriousness of trace element deficiencies is such that if some of us want to use more than 10-36 percent of our brain's potential we will have to make sure we're getting enough electrolytes—or we might not be able to.

By the way, it wasn't Einstein's observation. I looked it up. No one seemed to be sure who said it. This is a little quiz to see if you remember what I was talking about at the beginning of the chapter...

# ANTIOXIDANTS AND IMMUNITY

*"...the cost of a thing is the amount of what I will call life, which is required to be exchanged for it, immediately or in the long run."*

❦

Henry David Thoreau.

## CELL PROTECTION...

A number of the 'life enhancing' innovations of the last hundred years are having a devastating effect on our bodies. They may enhance our lifestyle, but they're not enhancing our *lifespan*. We know some specifics of how they harm us, and can only guess at others, but we do know that the bottom-line is the production of free radicals in the body and the harm they do to cells.

All forms of ionizing radiation, pesticides, heavy metals, asbestos, pollution, rancid and heated (particularly unsaturated) fats, burned, irradiated or moldy foods, heated proteins, and even stress load the body with highly toxic molecules called free radicals.

These molecules are enthusiastic, active little critters with a piece missing. Their main purpose is to seek out another molecule to join with in order to complete themselves. In the process they either damage or destroy the other molecule, or with it, form a new and highly toxic molecule.

Free radicals are not all bad, they are produced as a component of the immune system's cancer, bacteria and virus eradicating department, and they are used in the synthesis of vitamin D. Free radicals are also produced as a by-product of metabolism—when the cells burn sugar to produce energy.

If the 'foreign' free radicals or the body's own production of free radicals gets out of control, and its deactivating and recycling mechanisms can't keep up, then the actions of these mutant molecules damage cell membranes, enzymes and the genetic material in the cell nucleus. This is known as oxidation. When cells' genetic material, or the reproductive blueprint (RNA/DNA), is damaged they may divide uncontrollably and this, in the worst scenario, leads to abnormal cell reproduction and the growth of non-malignant, or cancerous tumors.

Free radicals are involved in the initiation of other degenerative disorders including cardiovascular disease, the deterioration of neurotransmitters (unsaturated fatty acids in the brain are highly susceptible to oxidation by lipid peroxide free radicals), impaired immunity and general premature aging. These are the molecules famous for cross linking skin tissue. Some forms of arthritis are a result of free radicals upsetting the lubricating quality of synovial fluid (joint oil) by oxidizing the lipids or fats in the fluid, destroying its lubricating effect and causing inflammation.

One of the two primary ways free radicals initiate disease and aging is by turning fats in the body rancid. This is known as lipid peroxidation. The cells are particularly vulnerable because they are surrounded by a membrane of fatty tissue, the cell membrane. Oxidation, or free radical damage either hardens cell membranes so that nutrients pass through with difficulty, or not at all, or the cells are damaged to the point where cell fluid drains out and they collapse. This latter is apparent in aged, wrinkled skin.

## RADIATION AND FREE RADICALS

Frequencies from the electromagnetic spectrum have played a part in human existence since life began. Until we discovered X-rays and radio waves, and learned how to generate them, our only exposure to electric and magnetic currents came from the earth and the cosmos.

In 1896 Thomas Edison gave the first public display of X-rays using his Fluoroscope machine, and started a novelty which, for the next few decades, would lead to the exposure of tens of thousands of people to dangerous levels of artificial radiation.

Throughout the 1920's workers painted luminous numbers on watch dials with radioactive paint, radium water was sold as a health drink and schizophrenics were injected with radioactive materials. Beauty salons used X-rays to remove unwanted hair, women's ovaries were irradiated as a treatment for depression, and hundreds of commercial gimmicks appeared that were based on X-rays. Everyone jumped on the bandwagon, Superman had X-ray vision and joke X-ray merchandise even found a market. A British company manufactured women's 'X-ray proof' underwear, presumably as insurance should Superman unexpectedly appear.

Artificial X-ray radiation is now the most strictly controlled material in the world, but we are still surrounded by ionizing radiation. About 15 percent of our exposure comes from nuclear fallout, nuclear weapons testing and medical X-rays, and 85 percent from natural sources such as cosmic rays, solar rays, and decay products of certain rocks and soils[27]. Ionizing radiation causes damage in the body by smashing molecules to pieces and producing free radicals.

There is also speculation that non-ionizing radiation or frequencies from the other end of the electromagnetic spectrum[28] can harm us. We are not completely sure what kind of damage 'artificial' forms of electrical energy do to us, but they can disrupt the brain and body's natural electrical field. Every thought, emotion and action is initiated by an electrical impulse 'firing' a nerve cell.

High level exposure to geopathic zones; disturbances in the earth's natural electromagnetic field, or to artificial electrical fields may disturb these delicate signalling systems[29]. This results in a disruption of biological processes such as communication amongst cells, white blood cell activity and the function of the immune system.

Some frequencies and strengths of extremely low frequency waves ELFs, especially those in the 50-60 Hz range, (mains electricity and electrical appliances) can alter brain tissue's release of calcium; the catalyst of neural impulses.

Researchers have found that ELF fields encourage higher production of a particular enzyme in mouse, rat and human cancer cells. This indicates that ELFs could encourage cancerous growth, but not necessarily cause it.

It is my belief that trace minerals exert protection from both forms of radiation by strengthening the cells, protecting their RNA/DNA, and enabling the production of antioxidant enzymes. Could our increasing susceptiblity to skin cancer and sun damage be due primarily to a lack of trace minerals and decreased production of the body's own antioxidant enzymes? Maybe it is time to take some of the blame off the damaged ozone layer.

## STRESS AND FREE RADICALS

'Stress' is one of those wonderful, all encompassing words we learned to blame almost everything on in the last two decades. Now we've gone over threshold and tend to be skeptical whenever anyone mentions 'stress' as a cause of disease or ill health. But our reaction to stressors is what generates dis-ease in the body, by creating toxic by-products such as free radicals. Stress triggers and intensifies the effects of all the other causes of disease and aging, and accelerates the use and depletion of nutrients.

The mind and body's reaction to stressors propels a flood of adrenal hormones, including noradrenaline, into the bloodstream, with a resulting increase in the production of chemicals transmitting nerve impulses in the body. These transmitters are primarily electrolytes or trace minerals. This drains electrolyte reserves and compromises the body's ability to detoxify both the stress hormones, and the free radicals.

Trace minerals are needed for the production of the SOD factors, or the antioxidant enzymes that scavenge cell-damaging free radicals. The more stress we experience, the more electrolytes we need to replace those used up by increased neurotransmission and to produce the enzymes that neutralize free radicals. And so it goes.

The stress and trace mineral research carried out by researchers at the Universities in Bethesda, Michigan, and the USDA Human Nutrition Research Center in Maryland produced clinical evidence that periods of physical and psychological stress deplete trace minerals. They also showed that it can take 7 days or more for mineral levels to return to 'normal' when the stress is 'off.' But what happens when someone lives with the repercussion of stress without respite—their reaction to their job, finances, environment, people and so on, and they eat the standard American diet? Essentially, they degenerate prematurely. And feel like hell while it is happening.

## CELL PROTECTION—
## ANTIOXIDANTS, ELECTROLYTES

The body produces antioxidant enzymes to neutralize free radicals and protect the cells. Superoxide dismutase or SOD is one of the body's main antioxidant enzymes. It removes the superoxide free radical which is responsible for cell damaging reactions and disease including arthritis, and aging. The body can produce up to 5 million units of SOD and its partner catalase every day.

Peroxide radicals are neutralized by glutathione peroxidase, an enzyme consisting of the amino acid glutathione and the trace mineral selenium. Peroxide free radicals initiate heart and liver disease, premature aging and skin diseases including skin cancer, eczema, wrinkling, age spots, dermatitis and psoriasis. These free radicals impair liver function as they attack fatty tissue. The liver is close to 50 percent fatty tissue and is therefore very susceptible to damage. The body's cells are also covered in a fatty tissue—and peroxide radicals damage cell walls.

There are numerous other antioxidants made by the body and there are also antioxidant nutrients like vitamins E and C, beta carotene and the trace mineral selenium. These nutrients clean up any free radicals that make it past the antioxidant enzymes. But the key is to support the body's own production of antioxidant enzymes because they remove free radicals 3-10 times faster than the nutrient antioxidants. It makes far more sense to enable the body to produce its own.

## HEALTHY CELLS, HEALTHY BODY

The body has many defenses against bacteria and viruses. The mucous membranes in the nose and lungs are one of the first. If invading organisms manage to infiltrate these, and if not intercepted by the immune system at this point, those which reproduce in intracellular fluid head straight for weak cells, break-in and, if the pH is also disturbed, use the cell as a place to clone themselves by the thousands. As they multiply, they use nutrients and oxygen meant for the cells, and produce waste substances that are harmful to body tissues.

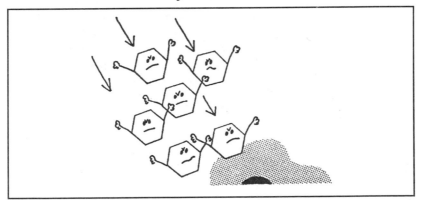

*Once inside the body, invading organisms take over cells that have been weakened by lack of electrolytes. Lowered osmotic pressure and altered pH of these cells makes it easy for a virus or bacteria to 'break in,' and reproduce.*

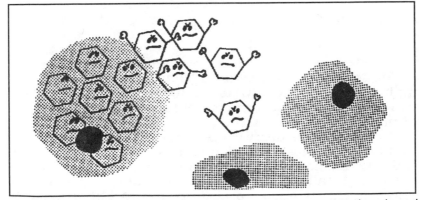

*As the viral organisms increase the cell bursts and they escape to invade and multiply in other weak cells.*

When cells are healthy—we are healthy. This is the key to health and longevity and protection against cancer, viruses, pollutants, chemicals and all forms of radiation. Keeping the cells, immune system, organs, elimination and the body's production of antioxidant nutrients at maximum capacity will enable us to survive under the assault of whatever the twentieth century throws at us.

Electrolytes are the primary factors for cell and immune system health. They keep cell membranes strong, raise their osmotic pressure, so that no virus or bacteria can enter, and maintain correct pH in intracellular fluid so that invading organisms cannot survive.

Healthy cells are kept round by osmotic pressure or osmotic equilibrium—an opposing fluid pressure between the minerals on the inside of the cell and the minerals on the outside of the cell. It is a critical balance and protects the autoimmune system in two specific ways:

① Bacteria or viruses cannot enter a healthy cell as the walls are strong and impenetrable.

② Trace minerals ensure that the pH of the cell is unfriendly to invading micro-organisms.

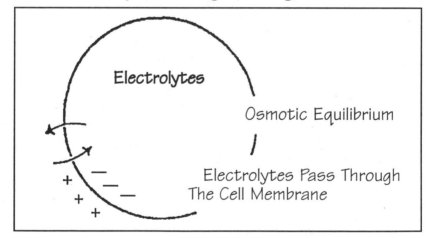

The body must receive an adequate mineral supply to maintain in each cell, what is known as osmotic equilibrium. Approximately 5 percent of each cell should consist of minerals.

According to Dr. Jerry Olarsch, "As we become deficient in trace minerals, the cell walls weaken and become irregular in shape, and their pH is upset. As the cell wall pressure breaks down, a virus can attack the weak spot and break in. If the cell's pH is unbalanced it becomes host to the virus, bacteria and so on, and this is the beginning of disease, whether it is a cold or cancer. Electrolyte balance is the critical factor. There is nothing else in nature which keeps cells round and strong. The body's pH and cell health, in addition to immune system strength, is the ultimate weapon against disease. This can only come through mineralization of the body. We have overlooked the most basic thing in our diet and as a result are running after a million cures. We can't see the forest for the trees."

When electrolyte balance is re-established in the body the cells become round and impenetrable. Re-establishing electrolyte balance has significant and enormous potential to prevent and eliminate disease.

Dr. Bernard Jensen writes, "It is time for scientists to realize that there is something deeper in man that has to be cared for. We cannot heal a person through symptomatic relief. We must go to the root of the problem—to the cell itself—where electrolytes are at work giving life. It is here that real healing can begin."

# BIBLIOGRAPHY

ALUMINUM/BRAIN

Constantinidis, J. Treatment of Alzheimer's Disease by Zinc Compounds. Drug Development Research, 1992; 27:1.14.

Martyn, C.N., et al. Geographical Relation Between Alzheimer's Disease and Aluminum in Drinking Water, 1989; 1:59-62.

Wenk, G.L., Stemmer, K.L. Sub-optimal Dietary Zinc Intake Increases Aluminum Accumulation in the Rat Brain. 1983; 288:393-5.

Ward, N.I., Mason, J.A. Neutron Activation Analyses Techniques for Identifying Elemental Status in Alzheimer's Disease. Journal of Radio Analyt. Nucl. Chem, 1987; 113(2):515-26.

Graves, A. B., et al. The Association Between Aluminum Containing Product and Alzheimer's Disease. Journal of Clinical Epidemiology, 1990; 43(1):35-44.

Science Watch. Minerals are Linked to Sleep Difficulties. The New York Times, Tuesday July 5, 1988.

CARDIOVASCULAR

Payar, L. Copper May Cut Cardiac Deaths. Medical Tribune, July 26, 1990.

Nebrand C, et al. Cardiovascular Mortality and Morbidity in Seven Counties in Sweden in Relation to Water Hardness and Geological Settings. European Heart Journal, 1992;13:721-727.

Singh R.B., et al. How Dietary Minerals Reduce Blood Lipids in Subjects With Risk Factors of Cardiovascular Disease. Trace Elements in Medicine, 1991; 8(1):29-33.

Can an Aspirin a Day Keep the Doctor Away? Rick Weiss. Health, May/June 1993

FREE RADICALS/IMMUNE SYSTEM

"Trace Elements and Immune Response," Editorial; Large Animal Vetinary Report, 1991;10(2):1:73

Trace elements in Human Health and Disease: Symposium Report. Environmental Health 26. World Health Organization. U.S. Environmental Protection Agency, 1987.

Willet, W.C. Selenium and Cancer. Lancet, July 16, 1983.

Ames, B.N. Dietary Carcinogens and Anticarcinogens, Oxygen Radicals and Degenerative Diseases. Science 221, 1983;1256-64.

Oberly, L.W., Oberly, T.D. Free Radicals Cancer and Aging. Aging Res, 1986; 8:325-71.

Cutler, Richard G., Ph.D. Antioxidants and Aging. American Journal of Clinical Nutrition, 1991; 53:373-379.

DiLuzio, N.R. The Employment of Antioxidants in the Prevention and Treatment of Experimentally Induced Liver Injury. Progressive Biochemical Pharmacological, 1967; 5:325-342.

Jesberger, James A., Dr. J.S. Richardson. Oxygen Free Radicals and Brain Dysfunction. International Journal of Epidemiology. 1991; 57:1-17.

Willet, W., et al. Blood Selenium Levels and Cancer Rates. Lancet, July 16, 1983;130-4.

Salonen, J., et al. Vitamin E and Selenium Blood levels and Cancer Incidence. British Medical Journal, Feb 9, 1985; 290:417-20.

Sies, H. M.D., et al. Oxidative Stress: From Basic Research to Clinical Application. The American Journal of Medicine. September 30, 1991; 91.

Shamberger, Baughman, Kalchert, Wills and Hoffman. Carcinogen-Induced Chromosomal Breakage Decreased by Antioxidants. Proceedings of the National Academy of Sciences. USA, May 5, 1983; 1461.

Beisel, W.R. Single Nutrient Effects On Immulogic Functions. JAMA, 1981; 245:53-58.

Chandra, R.K. Immunodeficiency in Undernutrition and Overnutrition. Nutrition Review, 1981; 39/6:225-231.

Gross, R.I., Newberne, P.L. Role of Nutrition in Immunologic Function. Physiological Reviews, Jan 1980; 60:188-302.

Mental and Elemental Nutrients, Carl Pfeiffer, Ph.D., M.D. Keats, 1975.

Nutrition Against Disease, Roger Williams, Bantam Books, New York. 1981.

OSTEOPOROSIS

Watts, David, D.C., Ph.D. Nutrient Inter-relationships Minerals-Vitamins-Endocrines. Journal of Orthomolecular Medicine, 1990; 5:11-19.

Reid, Ian R., M.D. et al. Effect of Calcium Supplementation on Bone Loss in Postmenopausal Women. New England Journal of Medicine, Feb 18, 1993; 328(7):460-464.

Dawson-Hughes B., M.D. et al. "A Controlled Trial on the Effect of Calcium Supplementation on Bone Density in Post Menopausal Women." New England Journal of Medicine NEJM, Sept 27, 1990.

Newcomer, A., Hodgson, S., McGill, D., et al. Lactase Deficiency: Prevalence in Osteoporosis. Ann Int Med, 1978; 89:218-220.

Osteoporosis. Journal of Bone Joint Surgery 1983; 65-A:274-278.

Mack, R., La Chance, P. Effects of Recumbancy and Space Flight on Bone Density. American Journal of Clinical Nutrition, 1967; 20:1194.

Keyler, D., Peterson, D. Oral Calcium Supplements: How Much of What, for Whom and Why? Postgraduate Medicine, 1985; 78:123-125.

Song, M., Adham N., Ament, M. A Possible Role of Zinc on the Intestinal Calcium Absorption Mechanisms. Nutrition Report International, 1985; 31;43-45.

Trace Elements Enhance Bone-Preserving Effect of Calcium Supplements. Geriatrics, July 1991; 46(7):67.

Rico, H. Alcohol and Bone Disease. Alcohol, 1990; 25(4):345-52.

Krall, E. A., Dawson-Hughes, B. Smoking and Bone Loss Among Postmenopausal Women. Journal of Bone Mineral Research, 1991; 6(4):331-8.

Li. K.C., Zernicke, R.F., Barnard, R.J. Effects of a High Fat-Sucrose Diet on Cortical Bone Morphology and Biomechanics. Calcification Tissue International, 1990; 47(5):308-13.

Gennari, C., Agnusdei, D. Calcitonin, Estrogens and the Bone. Journal of Steroid Biochemistry and Molecular Biology, 1990; 37(3):451-5.

RADIATION

Clarke, R. H., Southwood, T.R. Risks from Ionizing Radiation. Nature 338, pp197, 1989.

How Much Do a Few Hertz Hurt? Science and Technology, The Economist, April 16, 1988.

New York Power Lines Project Report: available from the New York Department of Health, Corning Tower, Room 2317, Empire State Plaza, Albany, NY.

WATER

Anderson, T.W. Water Hardness and Heart Disease. Journal of American Water Works Assoc., July 1976; 68:232-233

Burton, A.C., Cornhill, F. Correlation of Cancer Death Rates With Altitude and With the Quality of Water Supply of the 100 Largest Cities in the U.S. Journal of Toxicology and Environmental Health, 1977; 3:465-478.

Crawford, M.D.. et al. Cardiovascular Disease and the Mineral Content of Drinking Water. British Medical Bulletin, 1971; 27:21-24.

Crawford, M.D., et al. Changes in Water Hardness and Local Death-Rates. Lancet, August 1971; 327-329.

Pocock, S.J., et al. British Regional Heart Study: Geographic Variations in Cardiovascular Mortality and the Role of Water Quality. British Medical Journal, May 1980; 24:1243-1249.

Schroeder, H.A. Relation Between Mortality from Cardiovascular Disease and Treated Water Supplies. Journal of the American Medical Association, April 23, 1960; 98-104.

Schroeder, H.A. The Role of Trace Elements in Cardiovascular Diseases. Medical Clinics of North America, March 1974; 58:381-396.

Nebrand, C., et al. Cardiovascular Mortality and Morbidity in Seven Counties in Sweden in Relation to Water Hardness and Geological Setting. European Heart Journal, 1992;13:721-727.

## FOOD NUTRIENT LEVELS

Smith, B.L. "Organic Foods vs. Supermarket Foods: Element Levels." Journal of Applied Nutrition, 1993, volume 45 # 1.

Bear, Dr. Firman. "The Firman Bear Report." The Soil Science Society of America Proceedings, 1948.

# FURTHER READING

Trace Elements in Human and Animal Nutrition. 5th Edition, Volumes 1 and 2. Edited by Walter J. Mertz. Academic Press. (Editions 1-4 prepared by the late Dr. Eric J. Underwood).

Living Water. Olof Alexandersson. UK, Gateway books 1982. Distributed in the USA by The Great Tradition, 11270 Clayton Creek Road, Lower Lake, CA 95457.

The Body Electric. Robert O. Becker M.D., and Gary Selden. William Morrow & Co 1985.

Acres USA, Desk Reference Set.

Feed Your Kids Right. Dr. Lendon Smith.

Trace Elements, Hair Analysis and Nutrition. Richard A. Passwater, Ph.D., and Elmer M. Cranton M.D. Keats 1983.

Your Body's Many Cries for Water. F. Batmanghelidj M.D. Global Health Solutions Inc 1992.

Nutrition and Physical Degeneration. Weston A. Price D.D.S. Keats 1989.

Empty Harvest. Dr. Bernard Jensen. Avery 1990

Foods That Heal. Dr. Bernard Jensen. Avery 1988.

Nutrition and the Soil. Dr. Lionel James Picton O.B.E.. Devin-Adair 1949.

Soil Organic Matter. M. M. Kononova. Pergamon Press 1961.

Minerals and Your Health. Leonard Mervyn Ph.D., B.Sc., C.Chem., F.R.S.C. Unwin & George Allen 1980.

The Real Trace Element Problem. Michel Deville. Corbaz S.A. Montreux, Switzerland.

Fluid and Electrolyte Therapy. Donald E. Pickering M.D., Delbert A. Fisher M.D., Edward S. West Ph.D. Medical Research Foundation of Oregon 1959.

Micronutrient Interactions: Vitamins, Minerals and Hazardous Elements. O. A. Levander and Lorraine Cheng. New

York Academy of Sciences 1980.

Bread From Stones. Dr. Julius Hensel. Tri-State-Press. ISBN: 0-932298-85-0.

Complete Book of Minerals for Health. Sharon Faelten. Rodale Press 1981.

Mineral Balance in Eating for Health. Philip Chen Ph.D. Rodale Books Inc 1969.

Healthy Water for a Longer Life Martin Fox Ph.D. Dunaway Foundation, Amarillo, Texas.

Mental and Elemental Nutrients. Carl Pfeiffer Ph.D. M.D. Keats Publishing.

Secrets of Long Life. Dr Morton Walker. Devin Adair.

Secrets of the Soil. Peter Tompkins and Christopher Bird. The Perennial Library—Harper and Row.

Trace elements and Man. Henry Schroeder. Devin-Adair.

The Survival of Civilization. John D. Hamaker with Donald A. Weaver. Hamaker/Weaver Publishers.

Zinc and other Micronutrients. Carl Pfeiffer Ph.D. M.D. Pivot.

How to Raise Healthy Kids In Spite of Your Doctor. Dr. Robert Mendelson.

Minerals: Kill or Cure? Ruth Adams and Frank Murray. Larchmont 1974.

Acid and Alkaline. Herman Aihara. George Ohsawa Macrobiotic Foundation 1980.

# RESOURCES

**Electrolyte Formulas:**

Trace-Lyte™ liquid electrolyte trace minerals and Cal-Lyte™ liquid calcium with magnesium and electrolytes, send a stamped self addressed envelope for more information on Cal-Lyte™ and Trace-Lyte and other products.

Naturopathic Research Labs, Inc.
P.O. Box 7594
North Port, FL 34287
~~(813)~~ 426-5772

**Colema Board:**

The body begins to detoxify, usually within about 10 days, of using Trace-Lyte™ electrolytes. Naturopathic physicians may recommend the use of an enema, or preferably the use of a colema board (as recommended by Dr. Bernard Jensen in his book, *Tissue Cleansing through Bowel Management*) as a more rapid removal system for the toxins that are being broken down and eliminated. There are several boards available; the one considered most hygienic, safe, comfortable and strong is *The Ultimate Cleansing Unit*, available from:

Ultimate Trends, Inc.
7835 South 13E
Sandy, Utah, 84092
Tel: (801) 566-3191
Orders Only: 800-745-3195

**Warm Mineral Springs:**

Minerals are taken into the body from food and water and can also be absorbed through the skin. An hour in warm mineral water enables the concentrated minerals to draw out toxins from the body and penetrate tissue, neutralizing acids and assisting the body's own healing process.
*Warm Mineral Springs* in Sarasota County, Florida (next to the city of North Port) has perhaps the highest mineral content of any body of water other than the Dead Sea. It originates from a 2,500 foot deep geological fault and is believed to be the 'fountain of youth' that the Spanish explorer, Ponce De Leon, was searching for in the early 1500's.

Contact:
Warm Mineral Springs
12200 San Servando Avenue
Warm Mineral Springs, FL 34287
Tel: (813) 426-1692

## Juicerator:

Acme Juicer Manufacturing Co.
Waring Products Division D.C.A.
New Hartford, CT 06057

## Magazines:

*Remineralize The Earth*
Publisher/Editor Joanna Campe
152 South Street
Northampton, MA 01060 (413) 586-4429

*Acres, U.S.A.*
10008 E. 60th Terrace
Kansas City, MO 64133
Telephone (816) 737-0064
Fax (816) 737-3346

*Spectrum*
The Holistic News Magazine
61 Dutile Road
Belmont, NH 03220-5252
Telephone (603) 528-4710

*Explore* (for health professionals)
*Explore More*
P.O. Box 1508
Mount Vernon, WA 98273
Orders Only: 800-845-7866
Telefax: (206) 424-6029
Telephone: (206) 424-6025

*Longevity*
1965 Broadway
New York, NY 10023

*Let's Live*
320 N. Larchmont Blvd
3rd Floor, P.O. Box 74908
Los Angeles, CA 90004
Tel: (213) 469-3901
Fax: (213) 469-9597

*Townsend Letter for Doctors*
911 Tyler Street
Port Townsend, WA 98368-6541
(206) 385-6021

*Health*
P.O. Box 56863
Boulder, CO 80322-6863
800-274-2522

*Natural Health Magazine*
P.O. Box 57320
Boulder, CO 80322-7320
Tel: (303) 447-9330

## Soil Remineralization Products:

Peak Minerals, Inc.
101 North Tejon
Suite 200
Colorado Springs, CO 80903
Telephone (719) 635-8000
Fax (719) 635-8938

# GLOSSARY

**ABSORPTION:** The process of absorption refers to the transport of nutrients across the lining of the intestine into the bloodstream or lymph.

**BIOTRANSMUTATION**: The change of a chemical or element to another while within a living organism.

**COLLOID**: A substance, which may be gas, liquid or solid dispersed in a continuous gas, liquid or solid medium. The particles consisting of very large molecules or large aggregates do not settle (or do so very slowly). The system is neither a solution or a suspension.

**CRYSTALLOID:** A substance, like a crystal, which forms a true solution and can pass through a semi permiable membrane.

**HOMEOSTASIS:** Equilibrium, a state of balance.

**IATROGENIC**: Disease induced in a patient by a doctor's actions or treatment.

**pH:** The body's acid-base balance (pH) is slightly alkaline. Fruit and vegetables (especially those grown organically on mineralized soil) are potentially alkaline-forming in the body due to their particular mineral content. Potentially acid forming foods include dairy products, seeds, meat and grains.

**PROTEINS/AMINO ACIDS:** Proteins are used for building and maintaining the body. Amino acids, the components of protein, are the 'building blocks' of life. There are many body proteins that have special functions including hemoglobin, plasma proteins, antibodies, hormones and enzymes. Proteins also help in the exchange of nutrients between cells, in the intracellular fluids, and between tissues and the blood and lymph. The trace elements enable the body to use proteins.

**TRANSMUTATION:** The change of one chemical or element into another.

# ENDNOTES

1. For more information contact: Citizen's for Health, P.O. Box 360, Tacoma, WA 98401 and enclose a S.A.S.E.

2. John Hamaker writes in his book *The Survival of Civilization,* that Justus von Liebig rescinded his claims later in life. He wrote this in the form of a letter which was published in the Encyclopaedia Britannic in 1899 (it was removed from subsequent editions).

3. Humans and animals need somewhere between 45 and 60 minerals for optimum health and longevity. Plants cannot create minerals, they can only take up what is in the soil, and the minerals are used to make amino acids, vitamins and essential amino acids.

4. Dr. T. M. Randolph's figures illustrate the average amount of minerals taken to grow 100 bushels of corn on an acre of soil: 125 lbs of potassium, 30 lbs of phosphorus, 50 lbs of magnesium, 20 lbs of sulphur, 2 lbs of iron, 3/5 lb of boron, 1/3 lb of manganese, a trace of copper, iodine and zinc.

5. Aside from the major impact on public health and insidious illnesses caused by pesticides, farm workers exposed to these chemicals are 6 times more likely than non-farmers to contract one type of cancer according to a study by the National Cancer Institute. 20,000 cases of cancer a year can be linked to pesticide use according to data from the National Academy of Sciences in 1987.

6. Colloidal humus particles (which transfer minerals from the soil to root hairs) are negatively charged and attract positive elements such as calcium, potassium, sodium, magnesium, manganese, boron and iron. Sodium nitrate added to the soil on a regular basis causes a change in the humus particles as the excessive sodium ions eventually crowd out the other elements making them less available to plants. The humus becomes coated with sodium and inundates root hairs with the excess. Finally the plants are unable to pick up the minerals.

7. Many farmers are questioning the sanity of using chemicals which ruin their land, leach into ground water, affect their own health and contaminate the food they grow. But they are trapped in another catch-22. To pay for the chemicals, it is necessary to grow intense yields of crops, and in order to do this it appears necessary to use

chemicals. The transition from chemical to organic farming can be long, and fraught with problems, sometimes with disastrous financial consequences. However, a less financially precarious route could involve re-mineralizing the soil at the same time as manuring and cutting down the amount of chemical fertilizer used. This would encourage the re-population of soil bacteria, increase yield naturally, and begin re-establishing healthy soil and plant growth while allowing gradual reduction of chemical input until it could be phased out.

8. Acid rain was first noted in the 1950's in Europe and has steadily spread over the world until it is now found virtually everywhere. EPA's Norman Glass notes the danger to forests, plants and soil organisms: "The foliage is assaulted from above while the roots are starved and poisoned in the soil."

9. The content and availability of zinc in foods is influenced by processing and refining and other less obvious factors. The addition of EDTA and polyphosphate complexing agents to foods reduces the amount of zinc available for absorption, while a vegetarian or vegan diet may be high in phytic acid which decreases the body's absorption of zinc. Phytic acid is found mainly in grains, unleavened bread such as matzo, nuts, legumes, and some tubers.

10. Numerous authors have written about the 'dangers' of drinking spring or mineral water because they hypothesize that the minerals will form dangerous deposits in the body. As I dug deeper into the research to find the answer to this altercation it became clear that when minerals are dissolved by nature in unpolluted water they are present in a form the body *can* use.

11. The lining of the gastrointestinal tract is both electrically and ionically charged, and consequently positive and negative mineral ions joined together in liquid (electrolytes) are absorbed more easily than other forms.

12. 'Inorganic' minerals are those in their 'raw' state, not having been converted by plants, dissolved in water or altered to a more bioavailable form like those available in some supplements; citrates, orotates, aspartates and peptonates etc.

13. A Naturopath or holistic practitioner will suggest individual dietary amendments in addition to mineral/herbal/vitamin supplements. A general approach to diet

can include live and whole foods (including fresh vegetable juices) and avoidance of white flour and sugar, non-sun ripened citrus fruit and juices, alcohol, coffee, black tea and processed or chemical-ridden foods. In some cases abstinence from nightshade foods; potatoes, tomatoes, peppers, eggplant and tobacco, can also help to bring about some positive changes.

14. The prophylactic use of aspirin has not been shown to prevent heart attacks in men younger than 50, or healthy people. It does make the stomach bleed, increases the risk of brain hemorrhage and can lead to a number of serious side effects which may not be diagnosed as iatrogenic (drug induced) and consequently treated as another disease—with more drugs!

15. 'Hardness' is determined by the level of calcium, magnesium and trace minerals present in water. Obversely, the less calcium and magnesium the 'softer' the water.

16. Hormones are produced by the glands, including those in the brain. Glands can only be healthy when adequate trace minerals are absorbed by the body. Vitamins and protein are of major importance in glandular health, but they are made available to the body and catalysed by minerals.

17. Taking certain antacids as a calcium source works out infinitely more expensive than taking a balanced and absorbable form of calcium and magnesium with trace minerals. About half of the calcium in these antacids is also a carbonate carrier so they actually contain far less calcium than appears on the label.

18. The sodium present in bone is enmeshed in the crystals of the insoluble bone minerals and is not generally available as a reserve. The kind of sodium the body needs is not just table salt or sodium chloride (which most people get too much of) but the sodium obtained from vegetables and ripe fresh or dried fruit, milk, cream, oatmeal, yogurt, fresh fish, eggs, root vegetables, nuts and fruit juices.

19. Very low cal diets (around 500 calories a day), anorexia or excessive exercise lower estrogen levels, triggering the loss of menstruation and causing bone loss especially from the spine. If lack of menstruation continues for a year or more, there is significant loss of bone. Being over-weight is not a risk factor for osteoporosis as fat cells have the ability

to convert hormones secreted by the adrenal glands, into estrogen.

20. Chronic alcohol use increases acidosis in the body, and decreases calcium absorption by changing the way the liver processes vitamin $D_3$, it also causes decreased bone formation, increases the loss of minerals like zinc in the urine and upsets mineralization. Excessive, and mal-digested fats may form insoluble calcium soaps in the body causing calcium to pass through the system unabsorbed. Protein is important for calcium absorption and metabolism, it is an excess of meat as a protein source which becomes negative as meat increases acid in the body. A predominantly vegetarian diet which includes meat once or twice a week is not likely to cause problems.

21. Women past menopause absorb as little as 7 percent of their dietary calcium. Calcium citrate is also especially important for people who are not manufacturing normal levels of hydrochloric acid (HCl) in the stomach, i.e. people under stress, people who are deficient in electrolytes, the elderly etc. The calcium in calcium citrate is bound to fruit acids. It is readily dissolved in water and easy for the body to absorb. Studies confirm superior bioavailability over calcium carbonate.

22. Manganese is becoming deficient in foods because of soil exhaustion and farming practices which use lime on soil. Lime causes a chemical reaction that results in manganese being kept out of the crops while the phosphorus content is increased. Too much phosphorus interferes with calcium metabolism. A similar problem occurs with other minerals including boron. The high phosphate content of artificial fertilizers tends to depress the uptake of boron from the soil.

23. When using either the Pill or estrogen replacement therapy it is important to note that high doses of vitamin C, over 1000-1500 mg a day can increase the bioavailability of estrogen in the body—possibly increasing estrogen-related side effects.

24. A substance called $D_3$ is formed in the skin when it is exposed to ultraviolet light. The same vitamin is found in foods like fish liver oils. $D_3$ is inactive unless the liver converts it to calcidiol, the main form of vitamin D found circulating in the blood. Parathyroid hormone (PTH) stimulates the kidneys to convert calcidiol into calcitriol or activated vitamin D.

25. Adequate iron facilitates the blood's oxygen carrying capacity and the release of energy; promotes the development of certain chemicals in the brain which affect mood, mental attitude and performance; and encourages the formation of enzymes responsible for efficient muscle function.

26. Alzheimer's disease is characterized by a degeneration of nerve terminals in the brain. Neurofibrils (tiny conductive pathways) become tangled resulting in incorrect messages being transmitted. Dr. Daniel Perl was one of the first, in the late 1970's, to discover aluminum in the brains of Alzheimer's patients and in 1985 he connected high levels of aluminum in drinking water to different forms of senile dementia, including Alzheimer's and Parkinson's disease. Norway's Central Board of Statistics states that there is a direct link between aluminum in drinking water and dementia.

27. Radon and thoron come from the radioactive decay of uranium which is particularly prevalent in granite. It rises from soil and rocks, mixes with the air and is diluted in the atmosphere unless it seeps into a building, becomes trapped and concentrates in the internal environment. Radon decay results in the formation of minute particles which attach to dust or water particles and may be inhaled in air or steam, deposit in the lungs and emit radiation.

28. The distinguishing factor between ionising and non-ionising radiation is their frequency. Radioactive materials capable of causing damage to cell nuclei are known as ionising radiation and have frequencies over 3000,000,000 Hz. Examples of ionising radiation include X-rays, alpha, beta and gamma rays, radon, nuclear radiation and fall-out, cosmic rays, and ultra violet light which is on the border of the X-ray region. EMFs or non-ionising radiation includes radar, telephones, microwaves, radio, television and electrical transmission.

29. Electroencephalographs register the human brain operating at between 1 and 20 Hz, which is the same frequency as the natural electrical waves surrounding the earth—alternating low frequency resonances known as Schumann waves.

# INDEX

Insulin: 42, 45, 48–49, 55, 58, 61
Intestines: 28
Intracellular fluid: 72–73
Iodine: xv, 16, 86
Ionizing radiation: 68–69, 78
Iron: xiv–xv 13, 16, 22–23, 35, 40, 49, 56–58, 61, 86, 90
Iron peptonate: 22
Iroquois: 4
Ischemic heart disease: 34–35
Jensen, Dr. Bernard: xiii, vi, 7, 74, 80, 82
Joints: 22, 25, 59
Justa, Sister Mary: 14
Kidney stones: 41
Kidneys: 11–12, 17, 39, 89
Kirlian photography: vi
Lactase: 77
Lactic acid: 28
Lactose: 40
Laxatives: 16
Lead: 24, 65, 67
Lecithin: 30, 63
Libido: 25
Light: 38, 89–90
Lime: 11, 89
Lipids: 69, 75
Liver: 11, 17, 20, 28–29, 32, 38, 51–54, 61, 64, 71, 76, 89
Longevity: 6, 22, 26, 37, 73, 83, 86
Lungs: vii, 64, 72, 90
Lymph: 50, 85
Macrobiotic: 81
Magnesium: vi, xiv, 11, 13–17, 24–26, 34, 40–41, 43–45, 57, 61–62, 64–65, 82, 86, 88
Magnetic: vi, viii, 69
Malnutrition: 46
Manganese: xiv–xv, 13, 32, 39–40, 44, 49, 56, 67, 86, 89
Margarine: 29, 31
Meat: 33, 43, 57, 85, 89
Memory: 60–61, 63
Menopause: 37–39, 43–44, 89
Menstruation: 43, 58, 88
Mental confusion: 18
Mercury: 18
Mertz, Dr. Walter: v, xv
Metabolic: xiv, 17, 46–47, 53, 57
Micro-organisms: xv, 6–7, 73
Microbes: 7
Milk: 27, 29, 32, 37, 40, 54, 88
Molasses: 13
Molybdenum: xv, 13–14, 61
Mucous membranes: 72
Muscle: xi, 26, 43, 46–48, 50, 56–58, 66, 90
Muscle cramps: 26, 43
Muscle-building: 47
Nails: 43
Naturopath: v, 87
Naturopathy: 91
Nervous system: xiii, 62, 67
Neurological repair: 63

Neurons: 62
Neurotoxin: 64
Neurotransmitters: 19, 35, 48, 52, 62, 63, 68, 71
Nickel: xiv–xv
Nightshade: 88
Nitrates: 14
Nitrogen: 4, 14, 21
Nitrogen-fixing: 14
NLP: 55, 91
Non-ionizing radiation: 70
NPK: 4, 21
Nuclear radiation: 69, 90
Olarsch, Dr. Gerald: v, ix, 73
Orotates: 87
Orthomolecular: 12, 16, 40, 77
Osmosis: xiii, 23, 35
Osmotic equilibrium: 73
Osteoarthritis: 39, 41
Osteomalacia: 17
Osteoporosis: iv, 18, 37–45, 77, 88
Ovaries: 38, 44, 69
Over-farming: 4
Overweight: 36, 46, 52–53
Oxidation: 68–69
Oxygen: xiv, 57, 63, 72, 76, 90
Ozone: 70
Painkillers: 62
Palpitations: 43, 52
Pancreas: 12, 17, 52–53, 61
Papaya: 9
Parathormone: 38
Parathyroid: 44, 89
Parathyroid hormone: 89
Pasteurization: 27, 29, 54
Penicillins: 16
Peptonate: 22
Peroxidase: 71
Peroxidation: 69
Pesticides: 29, 53, 68, 86
Pfeiffer, Dr. Carl: v, 77, 81
Pfeiffer, Ehrenfried: 14
pH: vi, xi, xiii–xiv 7, 12, 16–17, 22, 28, 39–41, 45, 47, 64–65, 72–74, 76–77, 80–81, 85
Phosphatidyl choline: 63
Phytic Acid: 87
Pill, the: 89
Plaque: 23, 33, 56
Plasma: 35, 50, 85
PMS: 1, 26
Poison: 64
Potash: 11
Potassium: vi, xiii–xiv 4, 13, 15–17, 21, 41, 58, 86
Power lines: 78
Prednisone: 16
Pregnancy: xiv
Protein: xi, 14, 28, 32, 38–39, 44–45, 47, 50–52, 55, 57, 62–63, 85, 88–89
Protoplasm: 6
Psoriasis: 26, 71
PTH: 38, 89
Ptyalin: 28

This book was produced on the Apple Macintosh IIsi, using the software application Aldus PageMaker 5. Proof prints and final output were made on the Hewlett-Packard LaserJet 4M at 600 DPI. Main type families used are Bookman 11 pt. and Book Antiqua 11 pt. Cover: Davis Designs, P.O. Box 45, Campton, NH 03223

## *About the Author*

Gillian Martlew is a graduate of the Institut de Naturopathie du Quebec. She is a doctor of Naturopathy, author, free-lance journalist and NLP practitioner with an extensive private practice in New Hampshire. She is the author of numerous books on health which have been published around the world, and she writes for many major women's and health magazines including Cosmopolitan and Elle. She appears on national television and radio and ran a natural therapy phone-in on BBC radio. Dr. Martlew has been invited to speak at numerous lectures, across the U.S. and in Britain. She is a member of the New Hampshire Association of Naturopathic Physicians, The Society of Authors, The British Holistic Medical Association, The Natural Medicines Society and Iridologists International.